GREAT INVENTIONS

Geniuses and Gizmos: Innovation in Our Time

Special thanks to:

Bozena Bannett, Robert Dente, Gina Di Meglio, Brian Fellows, Anne-Michelle Gallero, Peter Harper, Suzanne Janso, Robert Marasco, Natalie McCrea, Mary Jane Rigoroso, Steven Sandonato, Michael Skinner, Vaune Trachtman, Cornelis Verwaal

Published by TIME Books

Time Inc.
1271 Avenue of the Americas
New York, NY 10020

First Edition

ISBN: 1-932273-03-4
Library of Congress Control Number: 2003104203

TIME Books is a trademark of Time Inc.

We welcome your comments and suggestions about TIME Books. Please write to us at:
TIME Books
Attention: Book Editors
PO Box 11016
Des Moines, IA 50336-1016

If you would like to order any of our hardcover Collector's Edition books, please call us at 1-800-327-6388 (Monday through Friday, 7 a.m.–8 p.m., or Saturday, 7 a.m.–6 p.m. Central time).

Soft Cover Picture Credits
Cover:
The Very Large Array radio-telescope complex in New Mexico

©Lester Lefkowitz

Inset photos:
CD: Steve Percival—SPL—Photo Researchers; Rocket: The Granger Collection; Motorcycle: AFP—Corbis;
Chip: James A. Sugar—Corbis; Model T: Culver Pictures; X ray: The Granger Collection; Plane: Culver Pictures

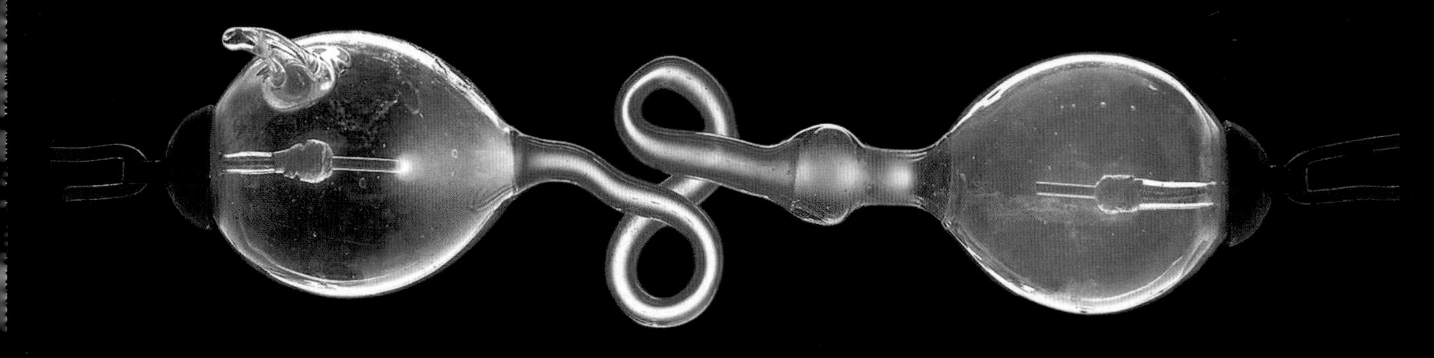

TIME GREAT INVENTIONS

Geniuses and Gizmos: Innovation in Our Time

Contents

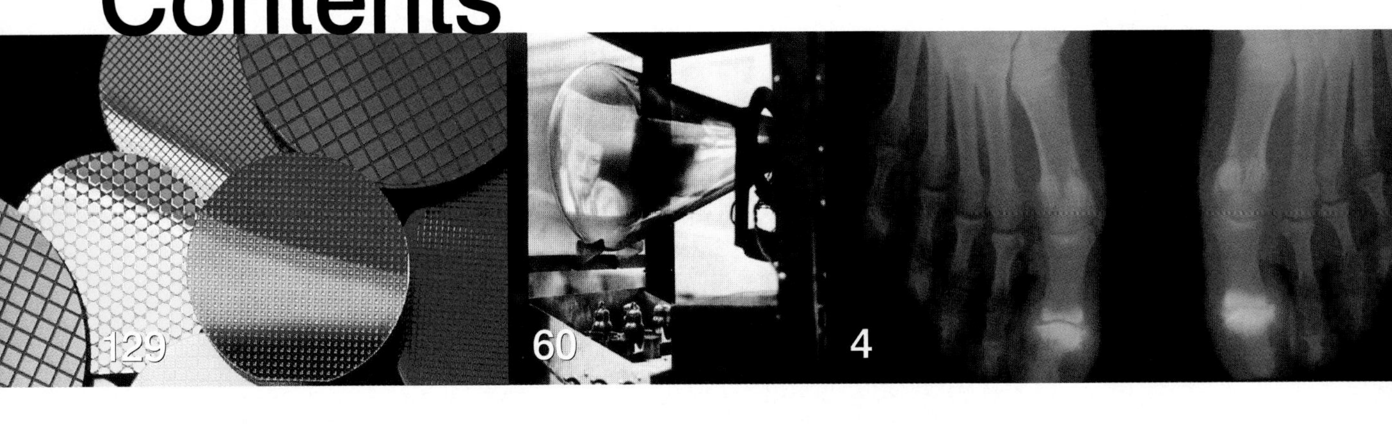

(LEFT TO RIGHT): STEPHEN GROHE—CORBIS; BETTMANN CORBIS; SPL—PHOTO RESEARCHERS; CULVER PICTURES (2); GEORGE SKADDING—TIMEPIX; OTTO ROGGE—CORBIS

115

733
27

161

142

Paying Attention to the

MANSELL COLLECTION-TIMEPIX

The dazzling parade of innovation marches on. But where's the drum major?

Not long into the deliriously enjoyable 1985 movie *Back to the Future,* Marty McFly (Michael J. Fox) knocks on the door of a sinister-looking house on a hill. The door opens—and there stands a figure who is instantly recognizable, though it's his first appearance in the film. It's Doc Brown, played by Christopher Lloyd. We know this guy, all right: with his unruly shock of white hair, his wide-eyed gaze, his manic intensity, he's the nutty professor, the mad genius, the crackpot, the crank. Sure enough, Doc Brown has concocted a crazy contraption—a time machine—the engine that drives the movie's plot.

Why begin a serious book about invention by invoking an absurd caricature? Because Lloyd's wigged-out nutcase is the single most memorable portrayal of an inventor in American culture in the past few decades. Looking for others? Good luck. We are no longer interested in the inventor as a type—even though inventions are still changing our lives. In the 1990s Internet revolution, the buzz was about the bucks, not the bytes. The brilliant folks behind the technology? Faceless hackers, they didn't merit our interest.

Yet the inventor figure is one of the most basic American archetypes. No, not the Gyro Gearloose of the cartoon, but his progenitors: the crafty New England peddler, the shade-tree mechanic, Mr. Fix-It, the Connecticut Yankee in King Arthur's Court. From the days when Americans first began to draw up a persona of our own—in the mid-18th century—the notion that we possessed a special way with technology was part of the blueprint.

The tintype, of course, was Ben Franklin. Much of this protean figure's legend flows from his mastery of science: as an ingenious inventor, he tossed off the Franklin stove and bifocals; as a serious scientist, he earned respect for his pioneering research into a new, mysterious phenomenon, electricity. His famous kite-flying episode, while foolhardy, made for terrific publicity—and for a terrific film scene, when it was re-created by Doc Brown in *Back to the Future.*

As America grew, inventors were among our most admired figures. Eli Whitney and Cyrus McCormick were benefactors of society, whose cotton gin and mechanical reaper earned farmers' blessings every day.

After the Civil War, a cascade of new inventions hurtled Americans into a new age: the telephone, the telegraph, the phonograph were wondrous devices, and each buzzed with the same magic juice, electricity, that Franklin first channeled with his kite. The colossus of the age was Thomas Edison, who was lionized and given the telling title the Wizard of Menlo Park. The wizardry, of course, was his mastery of electricity, an invisible,

MANSELL COLLECTION-TIMEPIX

Man Behind the Curtain

Don Herbert

vaguely threatening power. But everyone knew the plain, home-schooled Edison could be trusted to control it.

Soon, new Edisons arose: Henry Ford and the Wright brothers fit the native archetype: they were average folks (the Wrights owned a bicycle shop) who mastered gizmos that allowed Ford to defy time and distance, the Wrights to elude gravity.

In the 1920s, boys thrilled as they read about Tom Swift's nifty inventions; in the 1960s, the series was updated, and Tom Swift Jr. took over. In the 1950s, the gifted science popularizer Don Herbert was a staple of children's TV. As Mr. Wizard, Herbert still basked in Edison's incandescent glow.

But since the 1970s or so—where have all the inventors gone? Even our best popularizers—Bill Nye the Science Guy and National Public Radio's John Lienhard—do excellent work, but don't truly command the national stage.

Why the silence? When did inventors and their work cease to enthrall us? When did Ben Franklin turn into Doc Brown? And why?

I don't know the answers. But I have three lines of investigation to propose.

The first has become a cliché, but it can't be ignored: when the atom bomb dropped on Hiroshima, its mushroom cloud replaced the pedestal on which we had placed inventors. True, J. Robert Oppenheimer & Co. were more pure scientists than inventors, but they crafted an infernal device that still haunts our dreams—and Doc Brown's, whose corona of hair is a nod to Einstein.

Second, the advent of automation proved deeply dehumanizing. Ford's assembly line was brilliant, but it turned workers into automatons. The robot made his entrance into our imaginations in Karel Copek's play *R.U.R.* only eight years after Ford first automated his factory. Two decades later, Adolf Hitler's assembly-line death camps made automation an accessory to genocide.

Even more damaging, the process of invention itself became automated, bureaucratized. Edison didn't work alone: he had 500 colleagues at his "Invention Factory." But he was a great front man, and it was easy to ignore his elves. Take away Edison, and what do you have? Goodbye, Wizard; hello, R. and D. As a boy, TV pioneer Philo Farnsworth declared he'd grow up to be another Edison. Who today wakes up and says, "Some day I'll be a Du Pont chemist!"?

AMBLIN ENTERTAINMENT INC.

Back to the Future

Third, science receded from our grasp. Once every American could fix a Model T; it was a simple matter of mechanics. But can you fix your computer? Your microwave? Your cell phone? Today, technology seems to carry a warning label: FOR EXPERTS ONLY. Electronics? It has something to do with tubes, right? The digital world? Hey, let the hackers handle it.

Midway between the Wizard of Menlo Park and Doc Brown, there's another figure who sums up the decline of the inventor as an icon. It's the Wizard of Oz, and as he cranks his machine and the flames roar and the TV screen shows a scary electronic face, he thunders, "Pay no attention to the man behind the curtain." That's bad advice, it turns out, but we've followed it all too well. In these pages, though, we invite you to ignore the image of the crank for a change—and to pay heed to the minds behind the miracles.

—**Kelly Knauer**

DIGITAL VISION-GETTY

Anonymous

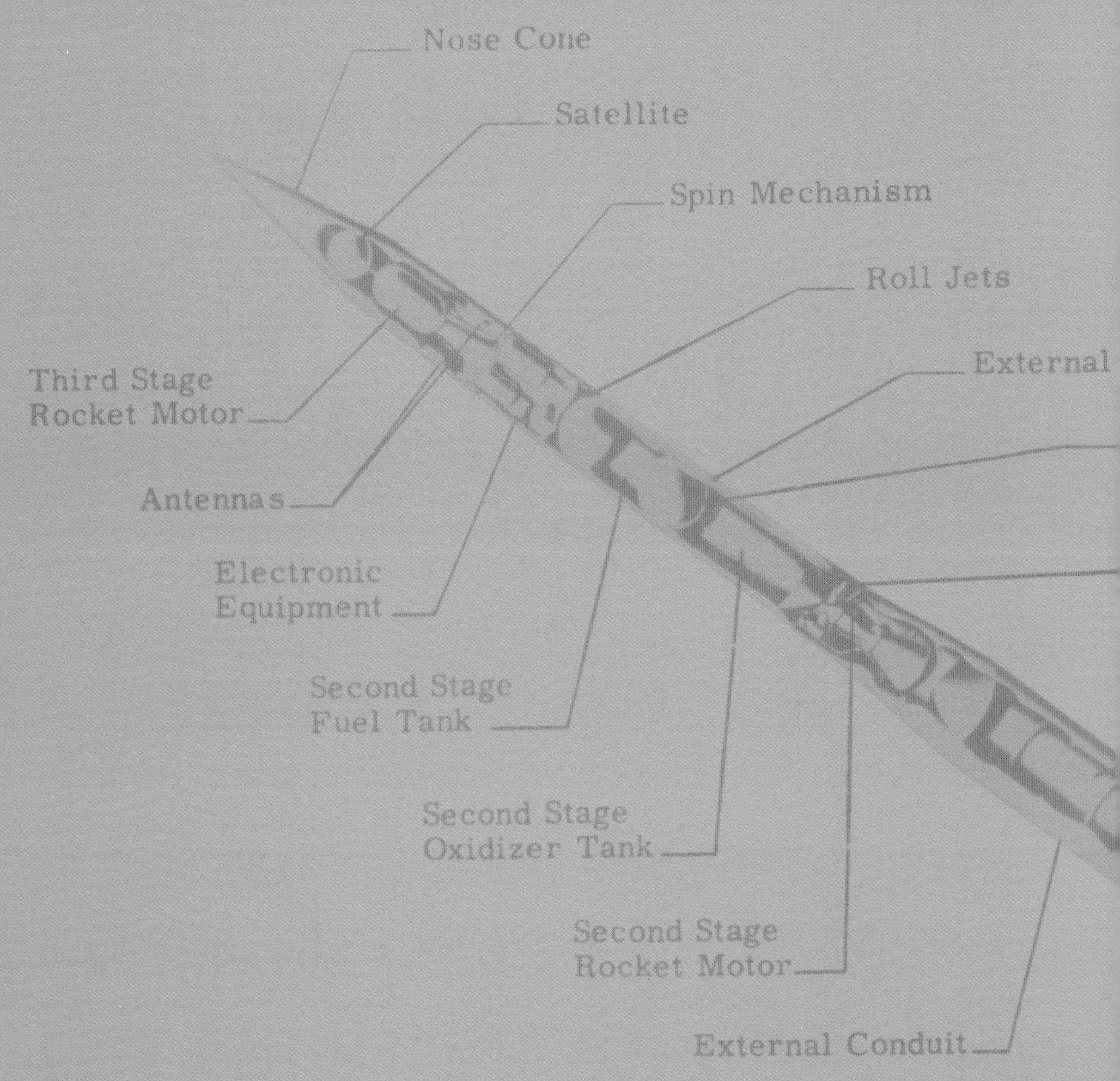

BETTMANN/CORBIS

How We Explore

duit

cond Stage
ium Sphere

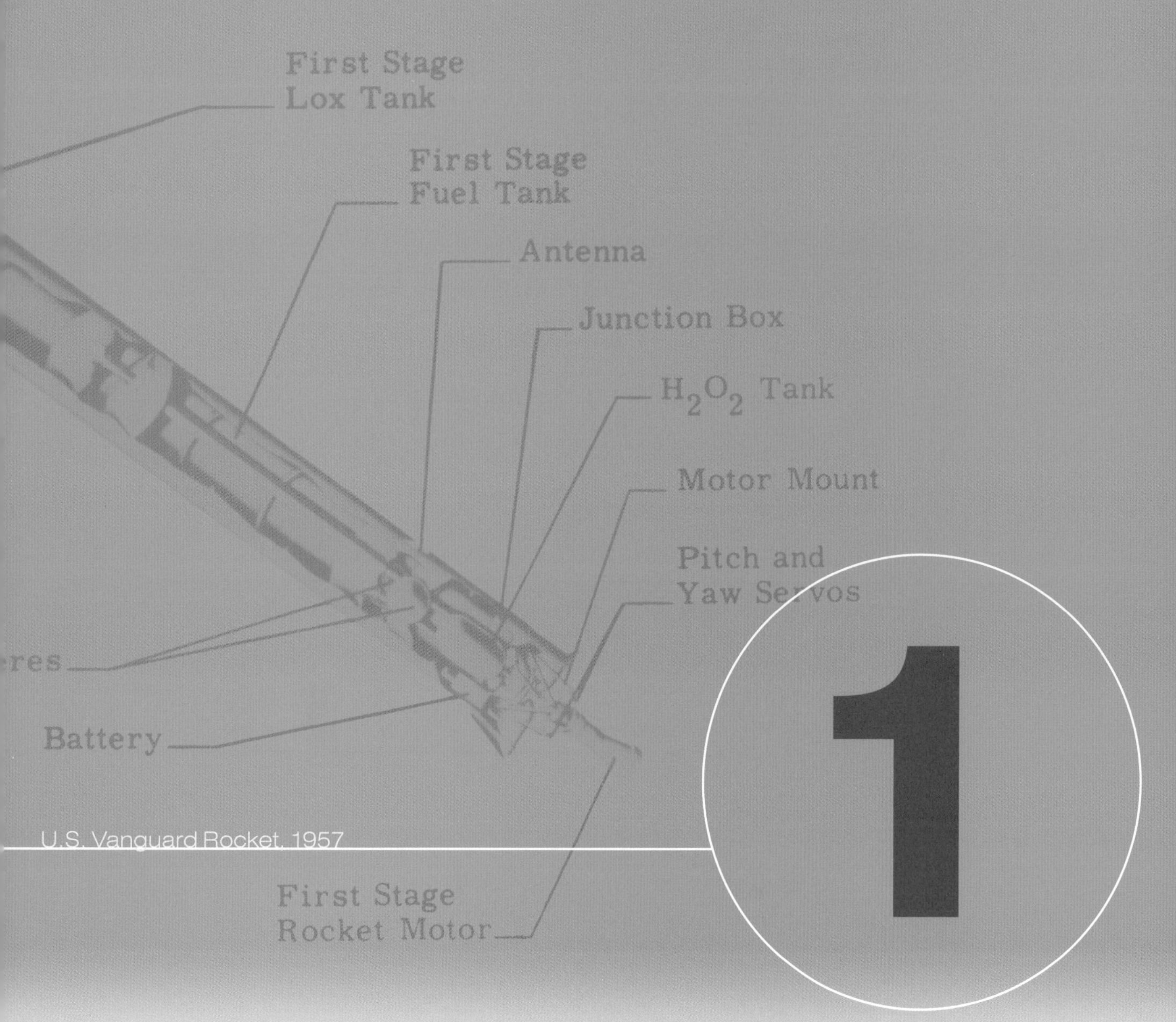

U.S. Vanguard Rocket, 1957

1

The Unseen World

Training new powers on nature, scientists amplified our vision

The late 19th century was a great age of invention, primarily powered by electricity. Yet even as Thomas Edison, Alexander Graham Bell and others were creating new ways to illuminate, communicate and record with electricity, research physicists were discovering even more recondite powers and harnessing them to examine the world in new ways.

One such mysterious force was observed for the first time on Nov. 8, 1895, by the German physicist Wilhelm Röntgen. He was working with the recently discovered cathode rays, which were created by running a strong electric current generated by an induction coil through low-pressure gases. To his surprise, Röntgen found that when he sealed the discharge tube in a carton that shut out light, he could make a barium-covered paper plate glow in the dark. Intrigued, he exposed various objects to the beams from the tube and recorded the results on a photographic plate. The objects seemed to become transparent. The next step: to direct the rays at a human being. Röntgen didn't look far for a subject: he had his wife place her hand on the photographic plate. In the resulting image, below, the rays rendered the flesh transparent but did not penetrate Mrs. Röntgen's wedding ring. The images—which the physicist initially feared might brand him a quack—were first called röntgenograms, but when the scientist named his discovery X rays (because of their unknown properties), the images themselves took the same name. Later research established that the rays were closely similar to light rays. Röntgen's work changed the course of medical history; he died of cancer in 1923. ■

CULVER PICTURES

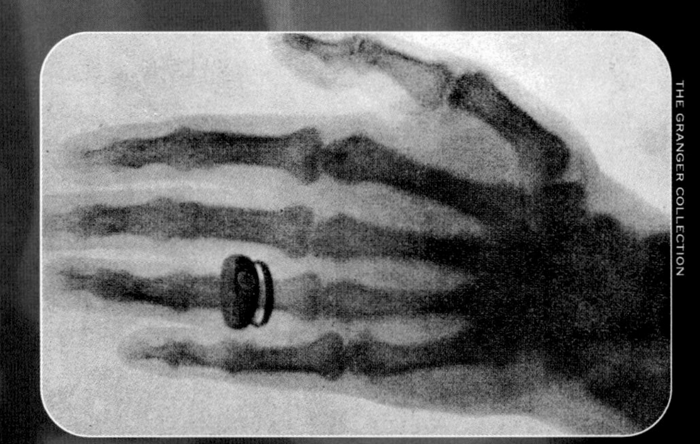

THE GRANGER COLLECTION

WILHELM RÖNTGEN
Unlike Albert Einstein, who worked in a realm of abstract thought, Röntgen, above, was a tinkerer who built his own apparatuses. He combined the best qualities of the inventor and the research scientist

SPL—PHOTO RESEARCHERS

COLIN CUTHBERT—SPL—PHOTO RESEARCHERS

Modern Microscopes

Though glass's ability to magnify objects had been known for centuries, the invention of the first true microscope is generally credited to the17th-century Dutch scientist Anton van Leeuwenhoek. Though optical technology constantly improved in the following centuries, the next great leap forward in microscopes did not occur until 1931, when German scientists Max Knott and Ernst Ruska invented the electron microscope, which beams accelerated electrons at a subject in a vacuum, allowing magnifications of up to 1 million times real size. Ruska was awarded the Nobel Prize in Physics for his work—55 years later! The critter below is a mosquito, as seen by a scanning electron microscope, above left.

A further evolution of the technology is the scanning tunneling microscope, which scans a subject's surface and provides a 3-D image of it. German Gerd Binnig and Swiss Heinrich Rohrer perfected the STM in 1981 and shared the 1986 Nobel with Ruska.

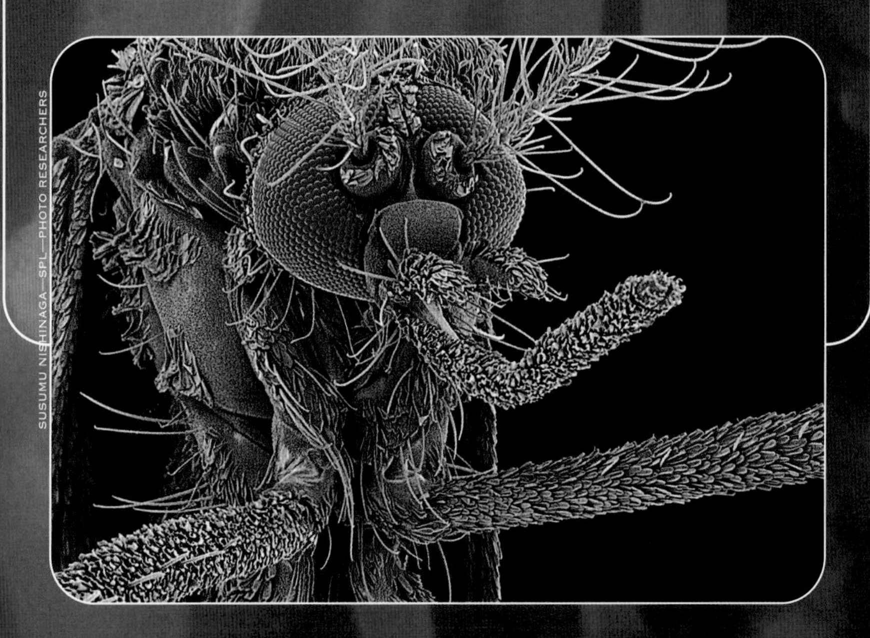

SUSUMU NISHINAGA—SPL—PHOTO RESEARCHERS

1947: CARBONDATING

U.S. chemist Willard Libby was first to show that organic beings take in small amounts of radioactive carbon while alive, which then decays at a steady rate. Finding the amount of carbon in a substance thus fixes its age. This skull, found in Chad in 2002, was carbon-dated at 7 million years old

PATRICK ROBERT—CORBIS SYGMA

1965: GPS

Like radar, the global positioning system, or GPS, was first designed for military use: the U.S. Navy pioneered the technology in 1965 aboard nuclear subs. The system works by beaming radio signals to stationary-orbit satellites, thus precisely triangulating the beamer's location. Quite an advance over Admiral Nelson's trusty sextant

THALES NAVIGATION

1935: RICHTER SCALE

Earthquakes like the monster that rocked Kobe, Japan, in 1995, below, used to be unquantifiable. The man who invented a system to measure their strength was Ohio-born Charles Richter, seen below at right with a seismograph. The modest professor refused to call his measurement the "Richter scale"

SHIGEO KOGURE—TIMEPIX

BETTMANN CORBIS

1955: ATOMIC CLOCK

A pair of British physicists, Louis Essen and Jack Parry, created the first atomic clock, below; it marked time by using a beam of atoms of cesium, a magnetic element, to measure the vibrations of a quartz oscillator. The result was so accurate that it might lose or gain one second in 300 years—yet the two later improved its precision

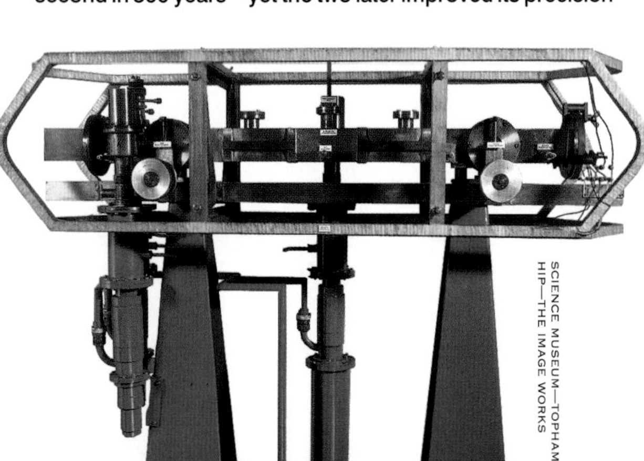

SCIENCE MUSEUM—TOPHAM-HIP—THE IMAGE WORKS

Secret Signals

Inventors dissect time, defy distance and quantify quakes

Predicted by British physicist James Clerk Maxwell in 1864, radio waves were harnessed for communication 30 years later and for detecting the unseen 40 years after that. Radio's brother is radar, an acronym for "radio detection and ranging," in which radio waves are beamed at an object, then bounced back to a receiving antenna. Radar waves were first successfully used by another Briton, Edward V. Appleton, who used them to measure the height of the earth's ionosphere. The first real working radar station was set up in 1935 by Robert Watson Watt; by 1939 a chain of secret stations circled the island—just in time to save it from German bombers in the Battle of Britain. ■

BOUNCE-BACK
An air-traffic controller uses radar to monitor traffic in Cincinnati, Ohio, in 1956. Radar, whose development was enormously accelerated by World War II, has turned out to have wide applications. It is used in geological exploration, and in its incarnation as Doppler radar, it is a life-saving detector of severe weather phenomena

1964: ALVIN SUBMERSIBLE

The *Alvin* is named for Woods Hole Oceanographic Institution scientist Allyn Vine, who first conceived it in the 1930s. The craft's titanium hull helps it withstand pressures at depths of up to 15,000 ft. A tethered vehicle, it broke loose from its cables in 1968 and spent 11 months on the ocean floor before it was raised

HULTON ARCHIVE—GETTY

Auguste Piccard

The Swiss-born adventurer was a throwback to overachieving 19th century explorers; he gained fame reaching the stratosphere in a balloon, then built a machine to take him deep under the waves. His "bathyscaphe," *Trieste,* was essentially a spherical cabin, not unlike Otis Barton's bathysphere, suspended beneath a buoyant gasoline-filled pontoon. Piccard's first dive, in 1953, reached a depth of 10,000 ft. in the Tyrrhenian Trench off southern Italy. Seven years later, with Auguste's son Jacques in command, *Trieste* submerged to 35,800 ft. in the Pacific's Mariana Trench—more than six miles below the surface.

HULTON ARCHIVE—GETTY

The Lower Depths

We journey ever deeper, but undersea science is still primitive

Underwater exploration on a modest scale dates back to the ancient world: the *Iliad* contains an account of an undersea dive. That's telling, for much of what we've learned about this cold, dark world has been driven not by scientists but rather by warriors seeking to create effective military submersibles.

Though pioneers like comet gazer Edmund Halley created primitive submarines, significant research of the undersea realm didn't begin until the 20th century. The first machine to explore the depths was the bathysphere designed by Otis Barton, an engineer who shared a fascination for the undersea world with William Beebe, an ornithologist. On Aug. 15, 1934, they made their first descent, in the Atlantic near Bermuda, reaching 3,028 ft., well over a half-mile beneath the surface. But it was a halting first step: their vessel was a tethered ball that could not move on its own. This first deep dive spurred interest in the seas, but with the world careening toward war, little was done to follow up.

The next great step in underwater research had to wait until after World War II. In fact, it was the work of a graduate of France's naval academy, Jacques Cousteau, and engineer Emile Gagnan. Their Aqua-Lung was a breakthrough whose by-products were more than scientific: the device allowed us to experience life in the depths for the first time, and its wonders created immense interest in this new frontier.

The trend got a boost from Auguste Piccard, whose untethered *Trieste* probed the deepest trenches of the ocean and pointed the way to a new generation of submersibles, including the Woods Hole Oceanographic Institute's famous diver, *Alvin*, which took its maiden voyage in 1964.

The latest submersibles are AUVs—autonomous underwater vehicles—untethered robots that can wander the depths for months. Like the exploration of the planets, tomorrow's undersea research may be performed by machines, which would amount to ... one giant step for robot-kind. ■

CULVER PICTURES

1943: AQUA-LUNG
Jacques Cousteau models his self-contained underwater breathing apparatus, or scuba gear. Cousteau almost perished on his first dive, not realizing that oxygen becomes toxic at depths below 30 ft.

WILLIAM BEEBE—NATIONAL GEOGRAPHIC SOCIETY

1934: BATHYSPHERE
William Beebe, left, and colleague Otis Barton, who invented the bathysphere after determining that a spherical shape was best suited to withstand the enormous pressures exerted by seawater at extreme depths. The bathysphere weighed 5,000 lbs.; the quartz-glass window was 3 in. thick; the cable that connected it to the surface and provided fresh air was 3,500 ft. long—472 ft. longer than their first dive

BROOKHAVEN ACCELERATOR

In the 1965 picture of the Brookhaven National Laboratory on Long Island, N.Y., below, we see a linear accelerator. Protons are fired through this "linac" at a target that is shielded from the operators by the concrete wall. The slower end of the 110-ft. "tank" is closest to us; particles speed up as theyapproach the target

The aerial picture at right shows the building of the Brookhaven "synchrotron" in 1958. Like a cyclotron, the 840-ft. ring moves particles in a spiral to accelerate them

BROOKHAVEN NATIONAL LABORATORY

In Pursuit of Particles

In order to decipher the atom, scientists had to smash it—safely

AP/WIDE WORLD

1931: CYCLOTRON
Lawrence and a cyclotron section. Today's accelerators are far more powerful than Lawrence's early versions, but the circular form remains the standard model

By the dawn of the 20th century, physicists knew that matter consists of atoms and that atoms consist of protons, electrons and neutrons. But what did these three particles consist of? If they could split the atom, scientists believed, they could not only discover its secrets, but also begin to control them. Perhaps, by controlling this process of nuclear fission, they could even release the atom's tremendous energy. The atom must be split. But ... how?

When E. O. Lawrence, a brilliant young physicist at the University of California, Berkeley, began tackling this problem in 1929, the first attempts at splitting the atom had just begun, using a linear accelerator—a device that fired a proton down a straight track at very high speeds, where it would smash an atom into its components. But linear accelerators had hit a theoretical brick wall. They worked fine when smashing lighter atoms like hydrogen, but the sheer momentum needed to crack open a "heavy" element, such as uranium, demanded more volts of power than could safely be controlled.

Lawrence's idea: Instead of pushing protons in a straight line, why not run them around a circular track, using magnetic fields to boost the accelera-tion just slightly with each pass? In this way, great speed could be achieved with incremental doses of electric power, rather than through one massive application of current.

In 1929 Lawrence built his first "proton merry-go-round" (later, the "cyclotron"), using $25 worth of scav-enged parts, including a vacuum pump, a radio-frequency oscillator, an electromagnet—and a pot of sealing wax. Within months, his team had found several new elements and particles. By the mid-1930s, the cyclotron had provided the first empir-ical verification of Einstein's famed equation on mass and energy, $E=mc^2$.

E.O. Lawrence was awarded the Nobel Prize in physics in 1939. Most of the significant applications of the sub-atomic world achieved in the last 70 years—including the atom bomb and nuclear energy—are the products of the cyclotron and its offspring. One discovery, a new heavy element with 103 electrons found in 1961, was christened Lawrencium.

But the outcome of his work that must have mattered most to Lawrence was more personal. He realized early on that a cyclotron's focused beam of radiation could be used to kill tumors—indeed, today's nuclear medicine is a direct result of the advances made possible by the cyclotron. Lawrence and his brother, John, a physician and medical researcher, began to treat cancer victims with radiation emitted from a cyclotron. One of the first pa-tients they cured was their mother. ■

Atomic Alchemists

AP/WIDE WORLD

Inspired by Lawrence's cyclotron, Irish physicist Ernest Walton, right, and British physicist John Cockcroft, left, scoured Cambridge University in 1932 for used bicycle parts, modeling clay, cookie tins, sugar crates and glass tubes from old gasoline pumps to build their own particle accelerator, a hybrid of the linear design and Lawrence's circular model. Rather than trying to manipulate heavier elements, they aimed to use a beam of protons to split one atom of a light element, lithium, into two atoms of an even lighter element, helium. On April 14, 1932, they succeeded in "transmuting" lithium into helium. It was the first time man had created nuclear fission; when Enrico Fermi and others learned to control this process, the nuclear age was born.

CHARLES FENNELL

HEARING AIDS
Completed in 1980, the Very Large Array near Socorro, N.M., harnesses the power of 27 receivers operating in unison to create a single gigantic "ear" tuned to radio waves emanating from stars, galaxies and quasars. Just as optical scopes fight atmospheric twinkle, radio telescopes seek to correct for "scintillation," the distortion of radio waves caused by Earth's ionosphere. The VLA antennas are organized in a *Y* shape; each of them is 82 ft. in diameter, weighs 230 tons and moves on tracks ... slowly

Big Eyes (and Big Ears)

With new instruments in the works, the future of telescopes is looking up

Since Galileo Galilei trained the eyepiece of his handmade telescope on Jupiter in 1610 and became the first human to see its moons, astronomers have fought a persistent enemy: Earth's life-sustaining cocoon of atmosphere acts as a filter that obscures our view of the heavens. The famed twinkle of that little star is the bane of scientists, one they've fought it in two ways: by placing their telescopes high above sea level and by increasing their size.

The Hale Telescope, completed in 1948, incorporated both trends. It was located atop Mount Palomar, Calif., its 200-in.-diameter Pyrex-glass mirror mounted in a 1,000-ton rotating dome. But optical telescopes couldn't just keep getting bigger—and heavier.

The advent of space travel gave astronomers a new idea: placing a telescope beyond the atmosphere. For years, they anticipated the launch of the Hubble Space Telescope, which went into orbit in 1990. But its primary mirror, alas, was faulty; it took a 1993 shuttle repair mission to get it in focus. The results were magnificent: Hubble images are now part of mankind's permanent storehouse of wonder.

Meanwhile, a major discovery revealed the need for a different sort of scope. In 1933 Bell Laboratories engineer Karl Jansky trained a radio receiver on the heavens, seeking the source of static that was interrupting phone calls—and found that radio waves were traveling throughout space. The pursuit of radio astronomy called for telescopes more like large ears than large eyes. Engineer Grote Reber built the first radio telescope in 1937 in his back yard in Illinois. Today's radio scopes, like the Very Large Array in New Mexico, above, are just a bit more complex.

Looking ahead, NASA hopes to launch a second-generation space scope by 2010. And earthbound optical telescopes are improving, thanks to a new technology, adaptive optics, that uses computers to factor out atmospheric distortion. ■

STEVE NORTHUP—BLACK STAR

EDWIN HUBBLE

Father of the theory of the expanding universe, astronomer Edwin Hubble, left, eyeballs the sky through a big scope at Mount Palomar in 1949. America's orbiting observatory, named for Hubble, carries a number of instruments, including spectrometers that measure infrared and ultraviolet phenomena. But amateur stargazers are most familiar with the beautiful images of deep space taken by the craft's optical scopes, like this stunning photo of the object known as NGC-6543, which astronomers believe may show the last throes of a pair of dying stars

BIG BOWL

The radio telescope in Arecibo, P.R., is a single giant parabola. Completed in 1963, it is the brainchild of U.S. electrical engineer William E. Gordon. Built on a 20-acre site, the big dish is 1,000 ft. in diameter and is made up of 40,000 individual aluminum mesh panels. Among other major discoveries, Arecibo confirmed the existence of gravity waves and found the first evidence of planets beyond the solar system

Going Up?

The space age started up in Aunt Effie's backyard

The rocket's red glare? The Chinese knew all about it; they developed self-propelled fireworks centuries before Isaac Newton explained the operating principle in his Second Law of Motion. But rocket science didn't really get off the ground until the 1890s, when English inventor William Hale added fins and angled exhaust tubes to an army missile, causing it to spin as a bullet does when shot from a rifled gun barrel, and thus to become more stable and easier to guide. In 1903, Russian educator and rocket enthusiast Konstantin Tsiolkovsky published *The Exploration of Space with Reaction Propelled Devices*, becoming the first to suggest that rockets could be powered by liquefied hydrogen and oxygen far more efficiently than the solid fuels that had been used for centuries.

Although Tsiolkovsky never sent a rocket aloft, his work fascinated an American scientist, Robert H. Goddard, who had begun testing solid-fuel rockets in 1909 but switched to a liquid-fuel design in 1915, after reading Tsiolkovsky. It wasn't until 11 years later, though, that Goddard was able to make his liquid-fuel design work. On March 16, 1926, he launched a missile

BROWN BROTHERS

ROCKET MEN
Goddard, second from right, and assistants at a Roswell, N.M., test site on April 19, 1932. He moved his tests to the desert after neighbors in Massachusetts complained. Below left, a German V-2 rocket lifts off from Peenemunde, on Germany's Baltic coast

Echo 1

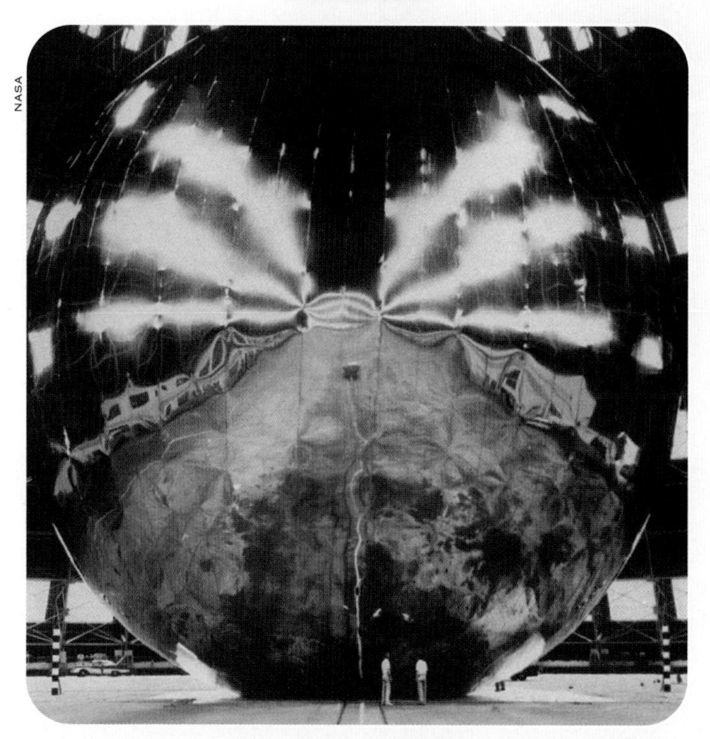

Of course it's not a rocket. But since it's a largely forgotten pioneer of space flight, we wanted to show Echo I. Launched in August 1960, the giant balloon, 100 ft. in diameter, was designed as a "passive satellite" that contained no instrumentation. Made of Mylar plastic dusted with vaporized aluminum, Echo I's highly reflective surface bounced electronic signals past the curvature of the earth and made possible the first coast-to-coast TV transmissions. Visible from the ground with the naked eye, Echo I inspired space-happy kids across America during its eight years in orbit. It was designed by North Dakota inventor Gilmore Schjeldahl (who also invented heat-sealed plastic packaging). The orbiting orb was recovered, semi-deflated, in May 1968, after years of pelting by meteoroids left its skin looking like a wrinkled prune.

that soared 41 ft. into the air before crashing back to earth—in his Aunt Effie's backyard in Auburn, Mass.

Goddard's flight is remembered alongside the Wright brothers' first takeoff as an epochal moment. He would be awarded more than 70 patents during the course of his career and achieve many additional historic firsts, such as his unprecedented use of gyroscopes to guide a rocket. But Goddard, like the Wright brothers, was hindered by a penchant for keeping his work secret. He also proved unable to integrate his brilliant ideas into a single operational system that would prove his detractors wrong.

For Goddard's work attracted derision in America: the New York *Times* ridiculed him in 1920, saying he lacked "the knowledge ladled out daily in high schools." But it was followed with religious zeal by rocket enthusiasts overseas, especially in Germany, where missile theorist Hermann Oberth founded the Verein fur Raumschiffahrt (Society for Space Travel). Among its early members was a young man who had decided at age 13 to dedicate his life to rocket science. By the mid-1930s, Wernher von Braun was able to talk the newly resurgent German military into subsidizing his experiments in rocketry. The ultimate result of his work was the V-2 rocket, which perfected many of Goddard's ideas—if perfected is the right term for a weapon so lethal to civilians.

Goddard died in 1945, just as rocket science was coming into its own. It was left to Von Braun, who emigrated to the U.S., to create the first multistage rockets and the mighty Saturn V that took Americans to the moon. But Goddard had the last word. In 1969, a few days after Neil Armstrong's historic step, the New York *Times* issued an official apology to Robert Goddard. ■

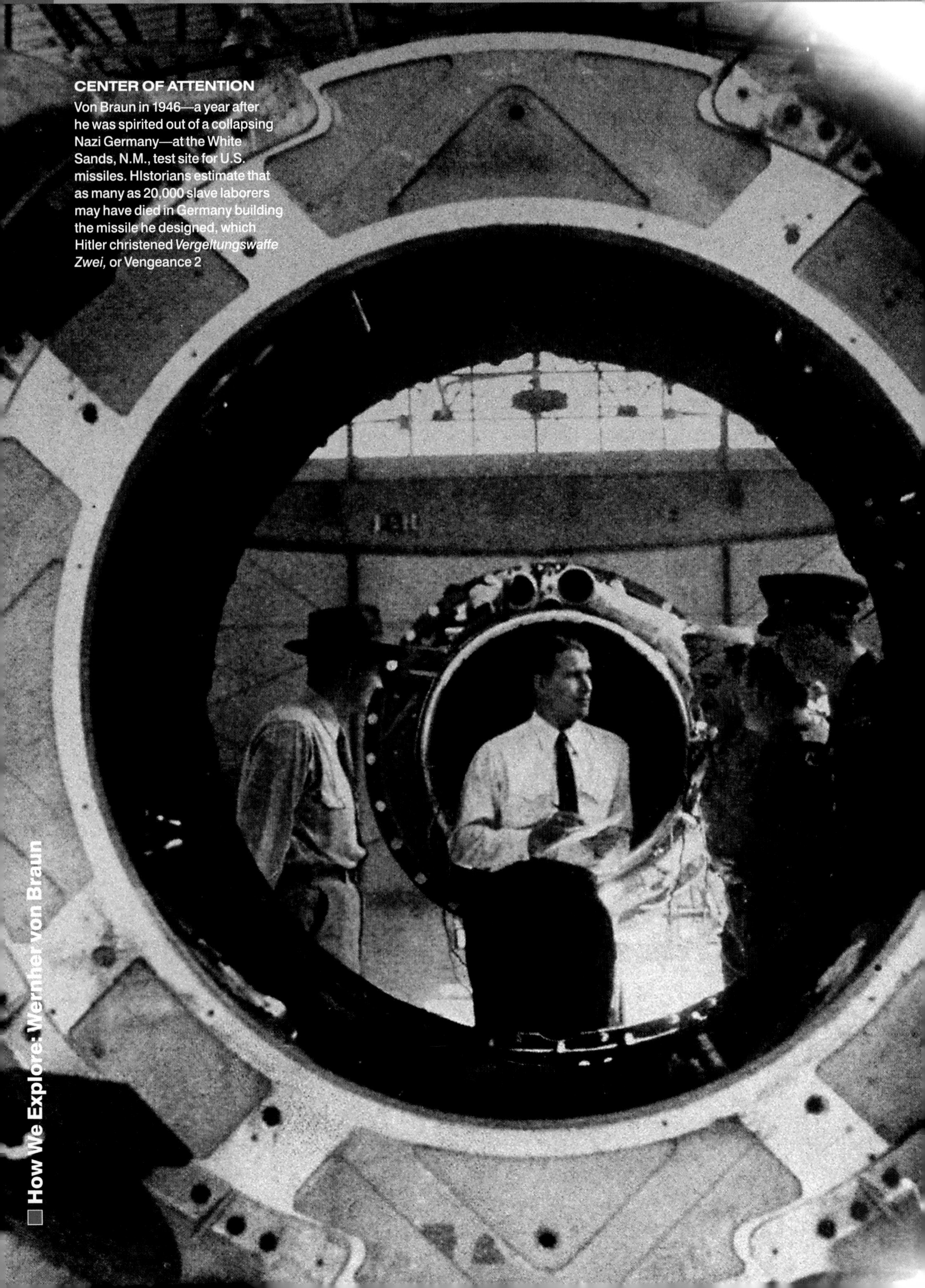

CENTER OF ATTENTION

Von Braun in 1946—a year after he was spirited out of a collapsing Nazi Germany—at the White Sands, N.M., test site for U.S. missiles. HIstorians estimate that as many as 20,000 slave laborers may have died in Germany building the missile he designed, which Hitler christened *Vergeltungswaffe Zwei,* or Vengeance 2

Profile

Blast from the Past

Wernher von Braun built Germany's V-2 missiles and America's Saturn V rocket. His patrons included Adolf Hitler—and Walt Disney

IN THE CLOSING DAYS OF WORLD WAR II, A YOUNG German bicycled down a Bavarian mountain road and breathlessly confronted a soldier of the U.S. Army. In broken English, the youth explained that his older brother, a very important scientist, was hiding in a ski chalet above, hoping to surrender to the Americans—before he was captured by the Russians. It wasn't until several officers questioned Magnus von Braun that the Americans realized who his brother was—Wernher von Braun, architect of the V-2 missile and the world's leading rocket scientist.

The Americans quickly hatched a secret plan, Operation Paperclip, by which Von Braun and more than 100 other scientists from the Nazi rocket program were spirited out of Germany, and brought to the U.S.

America needed Von Braun and his team, for it was clear that missiles—far faster than airplanes and almost impossible to shoot down—were the weapon of the future. And it was Von Braun who had come up with the two crucial insights that made missiles practical. He had adopted liquid fuel at a time when other rocket experimenters were fixated on traditional solid fuels, which at the time were both heavier and less powerful. And he had pioneered the use of gyroscopes and servomotors to stabilize top-heavy missiles, keeping them vertical after launch by increasing thrust on whichever side was listing off-course and then steering the rocket toward its target. These innovations added up to a 46-ft.-tall rocket that generated 55,000 lbs. of thrust at liftoff and could carry a one-ton bomb for almost 300 miles at speeds of more than 3,500 m.p.h. They also added up to more than 5,000 Britons killed by Von Braun's creation.

His innovations added up to a 46-ft. rocket with 55,000 lbs. of thrust. They also added up to more than 5,000 Britons killed

In America, Von Braun began upgrading U.S. military missiles with V-2 components and soon devised another breakthrough: 1950's Redstone multistage rocket, in which a portion of the launch vehicle fell away after its fuel was expended, making the remaining load lighter and easier to navigate. When the launch of Sputnik hurtled America into the space race in 1957, Von Braun claimed he could put a U.S. satellite into orbit with three weeks' notice. For political reasons, the Eisenhower Administration chose to use a nonmilitary rocket, the Vanguard, which Von Braun did not design, for the first mission. It exploded on liftoff on Dec. 6, 1957, before a live, nationwide TV audience. A month later, Von Braun's modified Redstone launched the Explorer I satellite into orbit.

Charismatic and charming, Von Braun entered the American mainstream; he even made films with Walt Disney to help promote space travel. The pace picked up: in 1960, he launched America's first intercontinental ballistic missile, the Atlas rocket. In 1961, Alan Shepard rode a reconfigured Redstone into space; a year later, a modified Atlas rocket lofted John Glenn into Earth orbit. When President Kennedy committed America to land a man on the moon, Von Braun left the military for NASA, where he built the giant Saturn V rocket. Its three stages stood more than 300 ft. tall and delivered 7.5 million lbs. of thrust at liftoff.

It was Von Braun's masterpiece. He wept openly when the astronauts who had ridden his vehicle took the first human steps on the moon. As the memory of his Nazi past dropped away and the glory of the Apollo mission kicked in, Wernher von Braun must have rejoiced: he had devised a multistage life. ■

THOMAS ... —TIMEPIX

First Steps to the Stars

Politics put space innovation on fast-forward; science reaped the results

It's easy to regard Robert Goddard's early rockets as inventions. But at first glance, the mighty machine at right, poised to haul the shuttle *Columbia* aloft on its maiden voyage, may seem too complex, or the product of too many hands, to be seen as an invention. Yet that's just what it is: Goddard's vision writ large. Satellites, shuttles and planetary probes are also inventions—or, more precisely, each is a congeries of inventions.

The first machine to travel beyond Earth's atmosphere launched the space age with a bang. The U.S.S.R.'s Sputnik was a primitive craft, about the size of a basketball. It weighed 183 lbs., carried a radio and not much more. That was enough. In 1957, at the height of the cold war, the notion of Soviet vessels plying the skies above America immediately sparked the creation of the National Aeronautics and Space Administration. The space race was on.

Driven by politics, prestige and occasionally even by science, invention in space has remained on overdrive ever since. Incredibly, less than 12 years after that first artificial moon beep-beeped its way around the globe, American Neil Armstrong stood on the lunar surface, his journey a triumph of innovation. Yet once the U.S. beat the Soviets to the moon, the drive into space lost its fascination for many Americans. And that's too bad, for the programs that have followed Apollo—including manned space stations like Skylab and Mir, planetary probes like Voyager and Magellan and complex satellites like Solar Max—are performing wonders of pure research.

Thanks to these machines of discovery, we have seen the surface of Mars and observed volcanos on Jupiter's moons. We have photographed the auroras of Saturn and located lost cities on Earth by tracing the long-buried tracks of ancient caravan routes. And, to the delight of adolescent boys everywhere, we have even discovered rings around Uranus. We have yet to launch a ship to reach the stars—but at TIME's invitation, two futurists took a crack at designing just such a vessel. The result—the *Starwisp*—can be seen on the following pages. ■

1981: SPACE SHUTTLE
Columbia prepares for lift-off on April 12, 1981. The craft broke apart during re-entry to the atmosphere 22 years later, on Feb. 1, 2003, killing all seven astronauts aboard

NASA

1973: SKYLAB
The first U.S. manned space station, Skylab was put together from spare parts of the Apollo program. Its cylindrical main unit was the "dry" third stage of a Saturn V rocket; astronauts traveled to it in Apollo command vehicles. Three crews of three astronauts occupied the Skylab workshop for a total of 171 days, conducting experiments in zero gravity. Skylab was planned to remain aloft for 8 to 10 years, but its orbit was affected by solar flares, and it burned up and scattered in pieces over the Indian Ocean in July 1979

1957: SPUTNIK II

The success of Sputnik 1, launched on Oct. 4, 1957, shocked Americans. But U.S. egos, already bruised, took another serious hit only a month later, when the Soviets launched hero space mutt Laika into orbit in a craft that was six times heavier. Re-entry? Forget it. Laika was put to sleep in orbit—or so went the story

1969: LUNAR LANDING MODULE

Unsung workhorse of the Apollo program, the lunar landing module, or LM, was used only six times, on missions dating from July 1969 to December 1972. It had 18 engines: two large rockets, one for descent to the moon and another for return to the Apollo's command module in lunar orbit, and 16 low-altitude control engines

NEXT STOP: THE STARS?

Engineer Robert Forward and theoretical physicist Freeman Dyson dreamed up this hypothetical interstellar voyager. Projected year of launch: 2503.

Microwaves beamed into space . . .

Carrying its own fuel will make Starwisp much too heavy to go fast; instead, it will get power from microwaves generated by a huge solar-powered satellite. In between launches, the satellite would send the power down to Earth to be sold

. . . and focused through a huge lens . . .

Made from rings of aluminum and floating far out in space, the lens will focus the microwaves on Starwisp; concentrated energy will fling the ship starward with 115 times the force of gravity

EARTH

MICROWAVES

SOLAR-POWERED SATELLITE To generate as much energy as Hoover Dam, its panels will be six miles (10 km) long

LENS It will be up to 30,000 miles (48,000 km) across—four times the size of earth

TIME Diagram by Joe Lertola
Source: Dr. Robert Forward

NASA

1977: VOYAGER 1

On Oct. 18, 1980, three years after its 1977 launch, Voyager 1 took this magnificent picture of Saturn. The planetary probe discovered three new moons of Saturn, as well as 1,000 new rings. Voyager is now the most distant man-made object in the universe

Solar System Missions

CORBIS SYGMA

1980: SOLAR MAX SATELLITE

The craft was built to study the sun during a cycle of intense sunspots (the solar max), but its mission was threatened nine months later when a simple electrical fuse blew and it could no longer point toward the sun. Three years later, a shuttle rescue mission repaired the problem, and the satellite went on to provide a wealth of data on the sun's workings

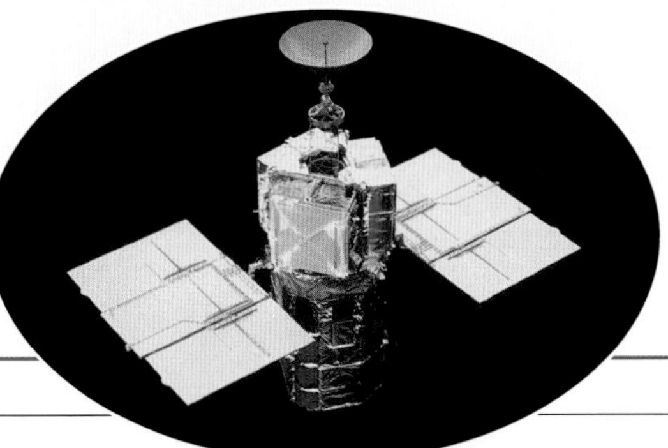

BETTMANN CORBIS

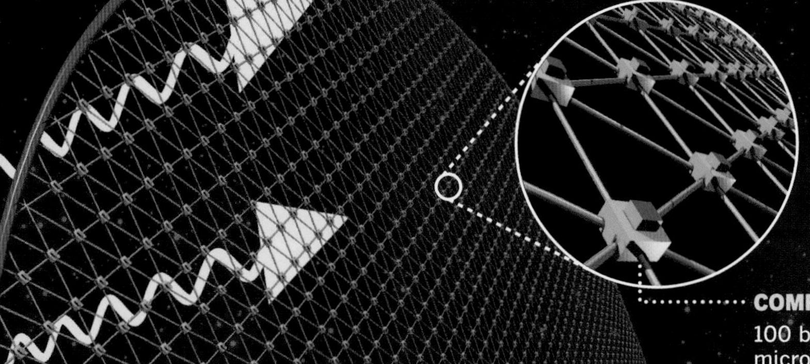

COMPUTER CHIPS
100 billion light-sensitive microprocessors linked by microscopic wires make Starwisp a combination camera, brain and transmitter

ALPHA CENTAURI

. . . may someday launch a spacecraft to the stars

After two weeks of microwave blasting, Starwisp, traveling at one-fifth the speed of light, would coast to Alpha Centauri in 21 years. Once it arrives, another short burst of microwaves would provide electricity to run the ship's microprocessors, letting Starwisp transmit images back home

STARWISP Although it spans four miles (6 km), Starwisp weighs only one ounce (28 g). To carry passengers, ship, lens and power station, it would have to be a million times bigger. It could be possible—in 500 years or so

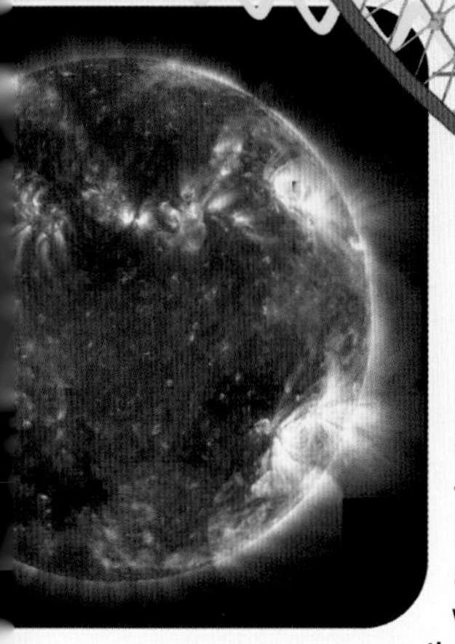

1998: TRACE SATELLITE

The TRACE craft (Transition Region and Coronal Explorer) was designed to study the sun during one of the star's periods of maximum flares, above, which affect our weather and communications. The 469-lb. satellite was lofted into its orbit 230 miles above Earth by a rocket dropped from a jet flying at 39,000 ft.

1997: SOJOURNER ROBOT

This planetary probe, only 1 ft. tall and 2 ft. long, landed on the surface of Mars on July 4, tucked inside a larger Pathfinder landing vehicle. It employed five laser beams to help detect obstructions like this rock, which NASA scientists playfully christened Yogi. Controlled from Earth, the probe moved slowly, only 2 ft. per minute. Sojourner conducted 20 chemical analyses of Martian soil and lasted three times its design lifetime of 30 days

NASA

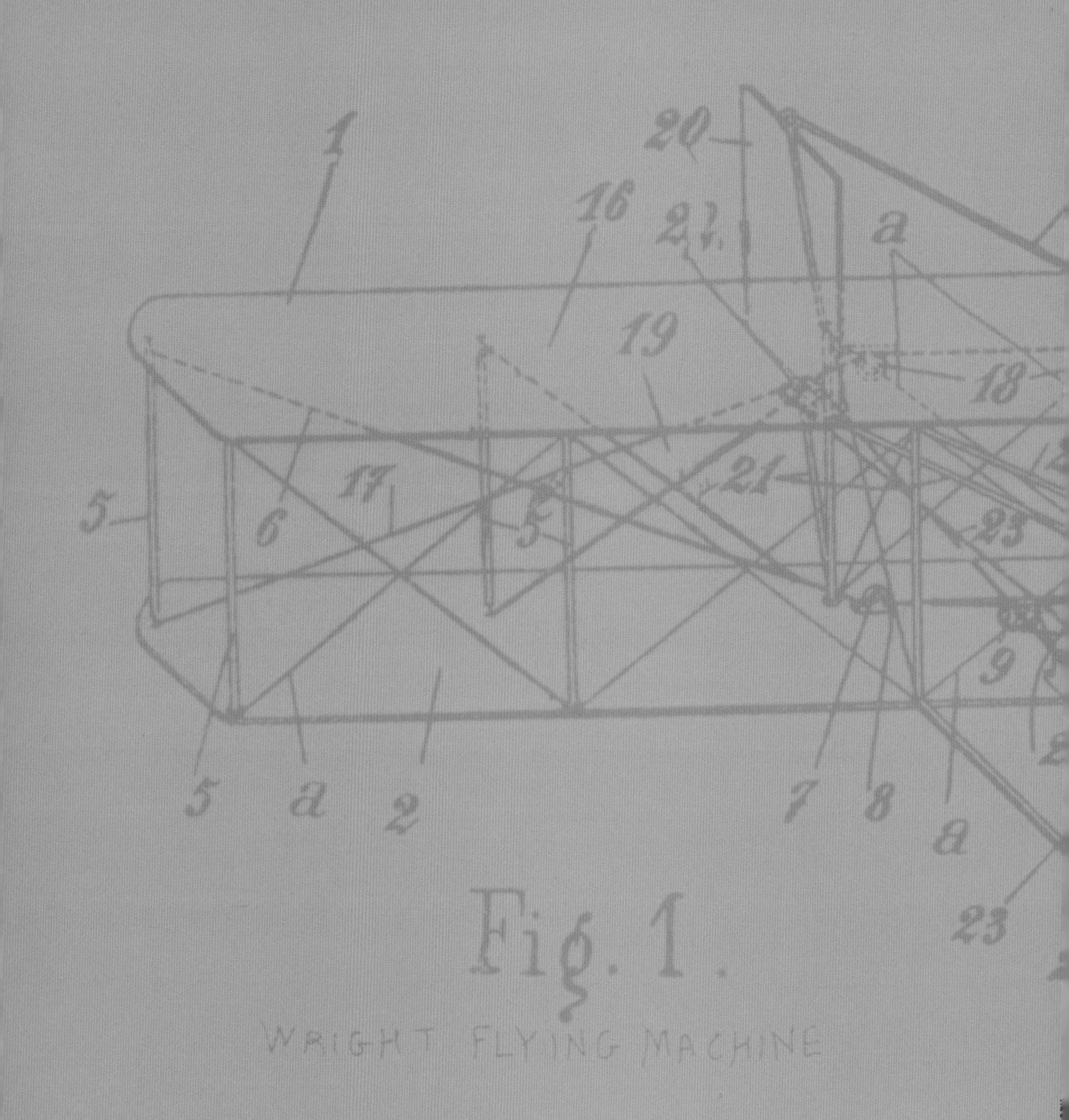

How We Move

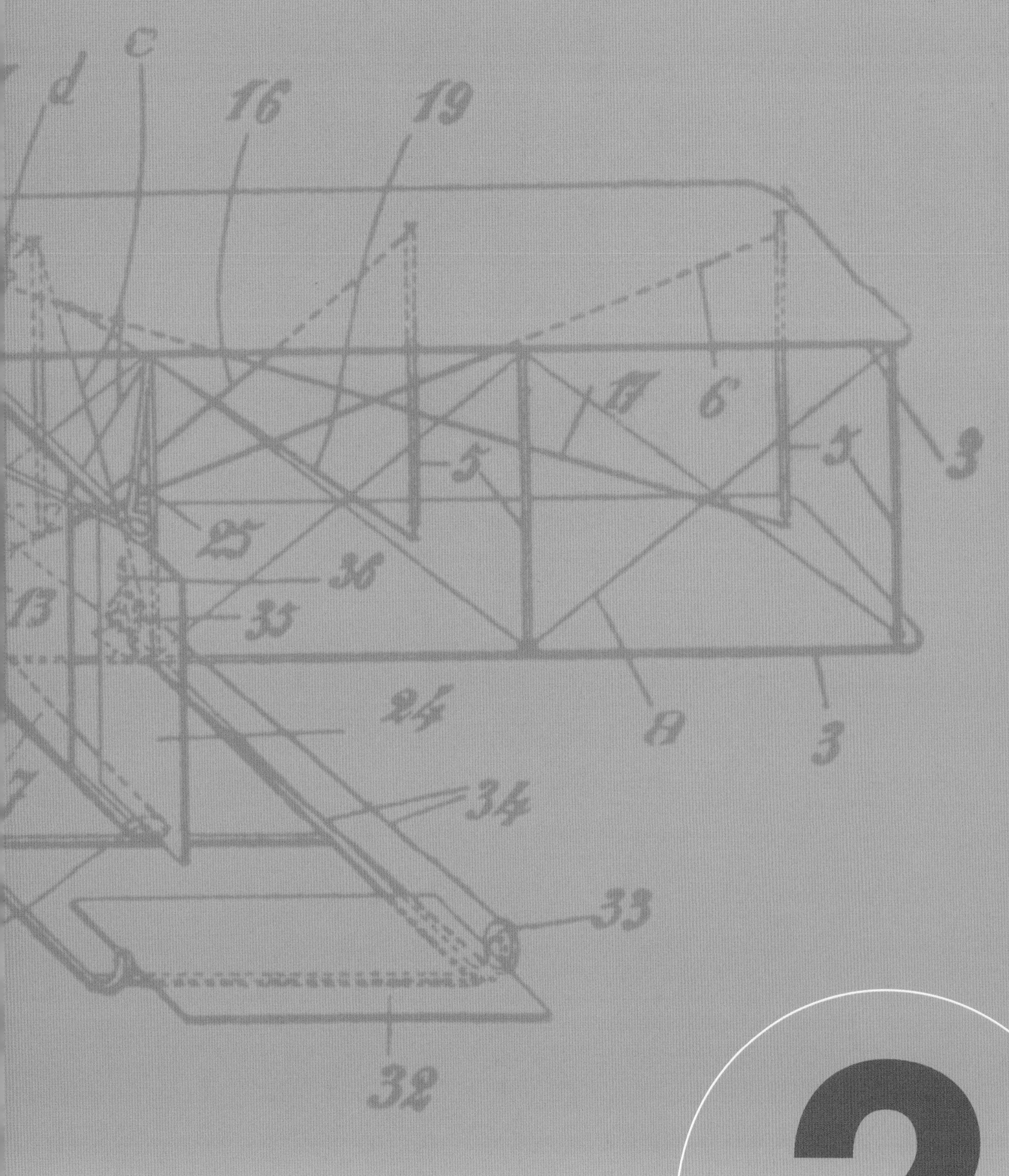

Wright Brothers Flyer, patent drawing, 1908

2

THEY'RE OFF!
Cyclists risked death atop their penny-farthing racers in 1890

Speed You Can Straddle

A Victorian oddity shifts gears, adds a motor and becomes a "crotch rocket"

"Be it known that I, Pierre Lallement of Paris, temporarily residing ... [in] Connecticut, have invented a new improvement in velocipedes." So begins U.S. patent No. 59915, issued in 1866, launching the age of the modern bicycle.

Earlier cycles—the celerifere, draisine, velocipede and others—relied on the rider to propel himself simply by pushing his feet along the ground. Lallement realized that pedals attached to the front wheel of these "bone shakers" would both propel the rider faster and improve his balance.

Soon, penny-farthing bikes hit the road—named for the largest and smallest coins of the time, resembling the bike's enormous front wheel and minute back one. The design breakthrough: the larger the wheel turned by the pedals, the farther (and faster) a cyclist would go with each revolution. The radius of the front wheel could be as large as the length of the cyclist's leg. Riders might be seated seven feet high, which made for nasty (even fatal) falls.

Even so, by the 1890s, bicycles were America's new, high-tech hobby. But Lallement had sold his patent for $1,000 two years after it was awarded. Alexander Pope, who eventually acquired it, made millions. Lallement, who later worked for Pope as a mechanic, died a pauper in 1891. ■

1886: THE SAFETY BIKE
Briton James K. Starley, who with William Hillman was a pioneer of the 1870 penny-farthing bike, made the big-wheelers obsolete with his "safety bicycle," father of today's bikes. Its key features: wheels of equal size, gears, solid rubber tires

State of the Art

1992: LOTUS SPORT

The process of inventing the bicycle continues. Recent designs (such as this swift two-wheeler, which set new speed records at the Barcelona Olympics) emphasize the use of lightweight, composite materials and improved aerodynamics, based on both the shape of the bike and the forward-leaning posture of the rider

THE IMAGE WORKS

HULTON ARCHIVE—GETTY

Evolution of the Motorcycle

Once bicycles become wildly popular in the 1880s, it was inevitable that someone would take the safety out of the "safety bike" by grafting it to the newfangled internal combustion engine. A Briton and a German took charge, and the motorcycle was born … to be wild.

1884: THE PETROL-CYCLE

Edward Butler shows off his gas-guzzling trike, above. A year later, Gottlieb Daimler built the first working two-wheeled version of a powered bicycle

1946: THE VESPA

After World War II, Enrico Piaggio, owner of a bombed-out Italian aircraft factory, recruited helicopter designer Corradino d'Ascanio to create a motorbike with small wheels, a step-through design and a body crafted of metallic airplane skin. High-revving, thin-waisted and wide in the back, it was christened the Vespa, Italian for wasp

AFP—CORBIS

2002: THE TOMAHAWK

Dodge's design concept needs four wheels to support its 500-h.p. engine. Its top speed (theoretically, 400 m.p.h.) will cost top dollar (a projected $250,000). This road monster may never go into production—but hog fans can dream, can't they?

Prime Movers

From Ford's Tin Lizzie to Olds' merry mobile, the machine that moved mankind had many fathers

Once the first engine had been attached to a bicycle, a great race began: to develop the motorized horseless carriage that everyone understood would transform the world. But before the four wheels of an automobile could be made as reliable as the four legs of a horse, these engines had to run consistently. This goal awaited the invention of the carburetor, which mixed vaporized gasoline with air in just the right proportion. No single inventor can claim full credit for the carburetor (as is true for the car as a whole), but two men get much of the glory. Austrian Siegfried Marcus' 1864 Kraftwagen was equipped with a gasoline engine and became the first auto to move under its own power. German Gottlieb Daimler produced a more efficient engine for his own self-powered vehicle in 1885, the same year archrival Karl Benz built his first car.

Combining the carburetor with Nikolaus Otto's new four-stroke engine, motorcars could now outlast and outpace horses. With the addition in 1911 of Charles Kettering's self-starter (eliminating the need for a hand crank), the age of the automobile rapidly got into gear.

Even as the young auto industry was learning how to make cars, it was also figuring out how to manufacture them in the dizzying numbers required for profitability. Although Henry Ford didn't quite invent the assembly line,

THE GRANGER COLLECTION

1908: FORD MODEL T

In 1913, five years after the first Tin Lizzies (above and at right) hit the road, Ford instituted the moving assembly line, splitting production into 84 distinct steps. Output soared, costs fell—and Ford lowered prices and still made more profit on each car

1901: OLDS RUNABOUT

Ranson E. Olds had been tinkering with self-propelled vehicles for 10 years before he made his first gasoline-powered car in 1896. In 1901 he became the first to mass-produce cars. Left, he sits atop one of his curved-dash Runabouts, briefly the most popular car in America

HULTON ARCHIVE—GETTY; DAIMLER; CULVER PICTURES

1889: DAIMLER AUTO

More interested in engines than cars, German Gottlieb Daimler designed one-of-a-kind vehicles for wealthy customers, like this buggy for the Sultan of Morocco. Yielding to consumer demand, he produced a line of cars for French customers that was named for the daughter of his sales agent in Nice—Mercedes

MANSELL COLLECTION—TIMEPIX

1886: BENZ AUTO

German Karl Benz was granted a patent for his three-wheeled "vehicle with gasoline engine" early in 1886. Although he never met his rival, Gottflieb Daimler, Benz merged his company into Daimler's in 1926 and formed the brand now known as Mercedes-Benz

1876: THE FOUR-STROKE ENGINE

Nikolaus Otto, a German, was not the first to build an internal combustion engine (French engineers managed that feat in 1859), but his four-stroke engine sparked the development of the motorcar. It divided internal combustion into four steps: drawing in fuel and air, compressing the mixture, igniting it and expelling the exhaust

SCIENCE MUSEUM, LONDON; TOPHAM—THE IMAGE WORKS

Dead End

MANSELL COLLECTION—TIMEPIX

1898: HEILMANN ELECTRIC AUTO

As early inventors sought a power source to turn their wheels, electricity and steam battled with petroleum to become the auto's prime mover. Dozens of companies, one backed by Thomas Edison, produced electric models, like this double-jointed contraption owned by Russian nobility. But with high production costs and poor battery technology, electric cars were outpaced by rapidly improving gasoline combustion engines

1957: ROTARY ENGINE

German inventor Felix Wankel designed an energy-saving combustion chamber that turned in a continuous circle. It first went on sale in the 1964 NSU Wankel Spider sports car, left. Too radical and too expensive, Wankel's design never really caught on

he did perfect a technique that would become the template for industry around the world. When Ford began production of the Model T in 1909, there were perhaps a few thousand cars in the entire world, most of them custom-made for wealthy gadget collectors. By the time Ford's "car for the masses" was discontinued in 1927, more than 15 million of them had been sold.

As consumers fell in love with cars, innovation raced ahead. In 1925, Francis Davis made the first effective power steering system. San Franciscan Richard Spikes perfected the automatic transmission in 1932. The invaluable tubeless tire was introduced when automobile production resumed after a World War II hiatus.

When U.S. automakers abandoned true innovation in favor of sleek design and annual style changes, serious auto experimentation moved abroad. In Europe, France's André Citroën and later Germany's Ferdinand Porsche (spurred on by Adolf Hitler's vision of a *Volkswagen*) created affordable cars for the people. In the 1960s, Japanese engineers created small, well-crafted machines that quickly swept the U.S. market.

Few inventions are as widely used, or as widely beloved, as the automobile. But as gas fumes spew daily from tens of millions of exhaust pipes, the need to find new fuels to power our cars is no longer a dream but a necessity. ■

1934: CITROËN "TYPE 7"

Former arms maker André Citroën, inset, aimed to build cars that the average Frenchman could afford. His "tin snail" seduced several generations of Europeans with its front-wheel drive (a major innovation), its low carriage and its superb road handling. It remained in production until 1957

2020 Vision

The future is hydrogen-powered, say the big car companies—it's finding a road map that's tough

GM

What kind of car will you drive in 2020? If corporate planners can be believed, you will be trading in your internal-combustion, greenhouse-gas-fuming petroleum engine for a zippy model powered by hydrogen and electricity.

General Motors is banking on a design it calls Hy-Wire, above, which takes its name from the hydrogen fuel cell that will power the car and the drive-by-wire technology that will allow drivers to operate it by hand controls, not foot pedals.

Look for the Hy-Wire in 2010, GM says. Until then, car experts see hybrid vehicles like the Honda Civic model, below, which runs on a combination of gasoline and electricity, as the stepping stone on the path to hydrogen-fuel-cell motors. (While promising to offer a cheap, renewable, pollution-free energy source someday, such cells are still very costly.) Honda and Toyota have offered hybrid models in showrooms since 2001; Ford, GM and Daimler-Chrysler are each planning to bring out hybrids in the 2004 model-year.

Purely electric cars seem to have run into a dead end. New models were tested by all the major auto companies in the '90s, but none of them found a way around twin roadblocks: they are hugely expensive (currently more than $100,000, on average), and they need to be recharged (a process that takes many hours) every 100 miles or so. ■

FROM TOP: JAMES NAZZ—CORBIS; STONE—GETTY; PHOTODISC—GETTY

Road Warriors

1898: SPARK PLUG
Oliver Joseph Lodge, a British scientist, invented the spark plug. He is perhaps better remembered as the first man to transmit a message by radio, in 1894

1923: STOP LIGHT
Garrett Augustus Morgan,an African-American inventor who also made the first gas mask, was inspired to adapt railroad signals for automotive use by the sight of a fatal car crash

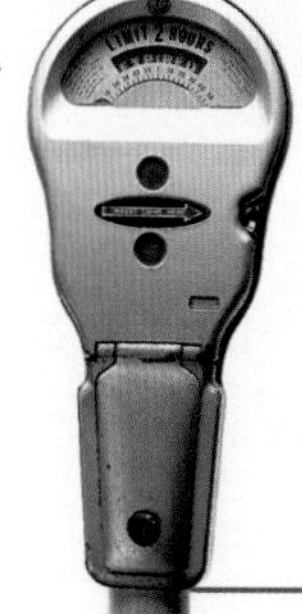

1935: METER
Oklahoma City newspaperman Carl C. Magee dreamed up the parking meter shortly after being named to the traffic committee of the city's Chamber of Commerce. Thanks!

INSTRUMENTS
Dashboard gauges alert you when the battery is recharging or assisting the engine

GAS TANK
With its 13.2-gal. tank and rating of 51 m.p.g. (highway), the Civic can cover up to 660 miles between fill-ups

The 2003 Honda Civic Hybrid

TIME Graphic by Lon Tweeten

GAS ENGINE
The 1.3-liter, four-cylinder engine is 25% smaller than a standard Civic's

ELECTRIC MOTOR
The 2-in.-wide, 13-h.p. motor is sandwiched between the engine and the transmission

BATTERY
The 63-lb. nickel-metal-hydride battery and controller power the electric motor

Spanning and Delving

Bridges are sublime elevations, geometry in repose. Tunnels, even in the building process, are boring

1883: BROOKLYN BRIDGE

Disabled by his father's creation—the pressurized caisson that allowed workers to descend beneath the East River and excavate the bedrock (as seen in the inset illustration)—Washington Roebling supervised the building of the Brooklyn Bridge from his bedroom window. He watched its progress through a telescope, and his wife relayed his written directions to the crew. In this picture, the bridge's double-arched pylons—massive yet graceful—are in place, but the roadway they hold up is not complete

FERNANDO ALDA—CORBIS

AFP—CORBIS

MODERN SPANS
Tension made visible, Calatrava's bridge in Seville doubles the drama by halving the number of pylons. Below, Eyre's brige in Newcastle is seen in "closed" position. To "open," the entire curved roadway lifts up—its arc matching that of the suspension arc— so ships may pass beneath

A suspension bridge is frozen physics, a basic exercise in hanging heavy structures from even heavier objects. The past 150 years of bridge building consist less of entirely new inventions than of refinements to existing techniques, a revolution in slow motion.

The story begins with John Roebling, whose suspension bridges at Niagara Falls, in 1855, and over the Ohio River at Cincinnati (then the world's longest, at 1,057 ft.), in 1866, foreshadowed the history-making work he would do in connecting Brooklyn with Manhattan decades later.

With every span, Roebling devised new methods to hold weight aloft with wire, stretching the limits of engineering (and the length of the span) a bit farther with each bridge. When designing the Brooklyn Bridge in the late 1860s, Roebling dared for the first time to use steel for the cable wiring. He also improved upon French inventor Marc Brunel's creation, adapting pressurized chambers or "caissons" to underwater work and using them to excavate the foundation of the bridge's two towers beneath the East River.

Roebling died before construction on the bridge even began, and his son Washington Roebling took over and saw the span to its completion in 1883. The younger Roebling did much of this work from bed, disabled on the job by "caisson's disease," a disorder caused by abrupt changes in air pressure, now called the bends. The finished bridge reached 3,460 ft. from shore to shore, making it the longest suspension bridge in the world and more than double the length of its nearest competitor.

Roeblings' modern counterparts are still in the suspension business, though their innovative designs may hide their debt. For the 1992 El Alamillo Bridge in Seville, Spain, architect Santiago Calatrava morphed the double-hung suspension bridge into the shape of a harp, with a single slanted pylon (more than 400 ft. high) gripping a 600-ft. span aloft with a graceful web of cables.

Wilkinson Eyre's 2000 Gateshead Millennium Bridge in Newcastle, England, is shaped like a giant eyelid. It is "closed" when its curved roadway is resting flat above the river (allowing autos and pedestrians to cross), but it blinks "open" when the entire roadbed pivots upward, with its raised arch allowing ship traffic to pass below. ■

Tunneling Technology

For centuries, tunnelers relied on only two tools: the hand pick and dangerously unstable black gunpowder. In 1846 Italian chemist Ascanio Sobrero first created nitroglycerin, which was far more powerful than powder yet still unstable. In 1869 Swede Alfred Nobel mixed nitroglycerin with inert minerals to render it stable until set off with a blasting cap. His new substance, "dynamite," revolutionized excavating.

CULVER PICTURES

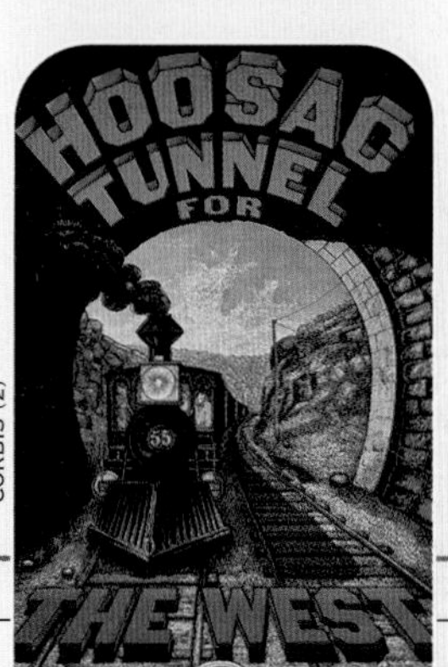

CORBIS (2)

Though Massachusetts' 4.7-mile Hoosac Tunnel took 23 years to complete (earning the tag "the great bore"), it was … ground-breaking. The world's first tunnel-boring machine was tested here, though it broke down after chewing through 10 ft. of rock. The Hoosac, which opened in 1875, also saw the first use of nitroglycerin in digging and the first use in America of a drill powered by compressed air, invented by Briton George Law in 1856.

Riding Motors to the Sky

Mastering power, lift and stability, a pair of bicycle mechanics taught us to fly

Daedulus' prodigal son Icarus failed. Leonardo da Vinci failed. A host of others could not conquer gravity. The two brothers who succeeded, Orville and Wilbur Wright, were Ohio bicycle mechanics who fiddled and sketched and flew kites and built wind tunnels until they finally cooked up a gadget that could fly.

Three crucial insights led the Wrights to victory in the race to the sky. First, most theories of flight at the time reduced the problem to lift (getting off the ground) and power (moving through the air). But the Wrights had a deep knowledge of balance from their bicycle work. They knew that a third component was just as important: stability. Many early aircraft actually created enough lift and power to get off the ground but once aloft soon capsized and crashed. Experimenting with kites and gliders, the Wright brothers developed a pulley system for altering the shape of a wing in mid-flight to change speed and point the nose of the aircraft up or down, much as flaps operate on the wings of modern airplanes. They also added a tail rudder to help steer.

Second, the Wrights re-thought the function of the propeller. Instead of using it to provide lift, as others had tried, their curved prop was used only to pull the plane

CULVER PICTURES

1903: WRIGHT FLYER

Orville, prone, mans the controls for history's first powered flight. Inset: Wilbur is on the right

MANSELL COLLECTION—TIMEPIX

Dead Ends: Early Attempts at Aviation

The two decades that straddled the turn of the 20th century were marked by great ferment in aviation experimentation, as wacky dreamers with goofy theories built ever more curious "flying machines." Some proved too fanciful, a few too far ahead of their time, and one was merely too late.

1906: DUMONT

Santos Dumont, a wealthy Brazilian living in Paris, became the first aviator to fly in public when he successfully piloted his 14-bis, a boxy biplane, before a French audience. Unfortunately for Dumont, he was three years after the Wright brothers' first flight, which was then unverified and still in dispute

MANSELL COLLECTION—TIMEPIX

MANSELL COLLECTION—TIMEPIX

forward. Horizontal speed, they found, was the key to achieving vertical lift.

Finally, the Wrights designed a curved, or cambered, wing to create that lift. They began by building a wind tunnel—one of the first such devices. Experimenting with moving air, they concluded that a fixed wing with a curved surface would be better able to create lift than the flapping, rotating and circular shapes others had attempted to use as wings. By driving air over the rounded surface on top of a wing, they forced it to move faster, and thus at a lower pressure, than the air on the wing's flat underside. The difference in pressure created lift. This breakthrough was the brothers' crowning achievement.

Their insights came together on the morning of Dec. 17, 1903, at a wind-swept beach in Kitty Hawk, N.C. While a startled local fisherman used Wilbur's camera to photograph the event, Orville flew a distance of 852 ft. (never much higher than shoulder level) in 57 seconds.

1896: LILIENTHAL
Mimicking the arched wings of birds, German Otto Lilienthal made some 2,000 "half-flight" glides from the tops of hills, learning the secrets of lift—before a single miscalculation of the wind sent him crashing to his death

MANSELL COLLECTION—TIMEPIX

1890S: "FLYING DOUGHNUT"
This gizmo appears to be more a precursor to the helicopter than the airplane. The picture was publicized in May 1927, as the race to be first to fly the Atlantic solo focused attention on flight. The winner: Charles Lindbergh

1900S: "FLYING WHEEL"
We don't know when this circular contraption was built. But we do know that it didn't fly. Vintage photos like this—with little supporting information—are all that remains of hundreds of failed attempts at flight that preceded the Wright brothers' success

CULVER PICTURES

For the next five years, while they awaited a patent, the Wrights refused to demonstrate their flyer in public, lest their ideas be stolen. Finally, in August 1908—patent in hand—they staged a public exhibition near Paris. As Wilbur executed a series of figure-eights above the crowd, spectators' hearts soared and skepticism vanished.

War (invention's father, if necessity is its mother) accelerated aircraft innovation. After World War I, aviation went through a giddy, barnstorming, adolescence. But soon passenger travel began driving airplane design.

In 1933 United Airlines took delivery of the Boeing 247. The first passenger airliner to feature wing flaps and retractable landing gear, the all-metal 247 could carry 10 passengers more than a thousand miles at speeds of up to 160 m.p.h. Alarmed, rival TWA turned to Santa Monica, Calif., aircraft genius Donald W. Douglas, who in the space of 30 months designed and built the Douglas Commercial series, the DC-1, DC-2 and DC-3. The first of these matched every performance benchmark of the Boeing 247; the second was so superior to the 247 that Boeing never sold another in the U.S.

But Douglas' greatest achievement was the DC-3. Able to carry 21 people much farther and faster than any plane before it, the DC-3 is the template for all modern airplanes. Several hundred are still flying today, airborne links between the age of the Wright brothers and the age of the right stuff. ■

Douglas DC-3

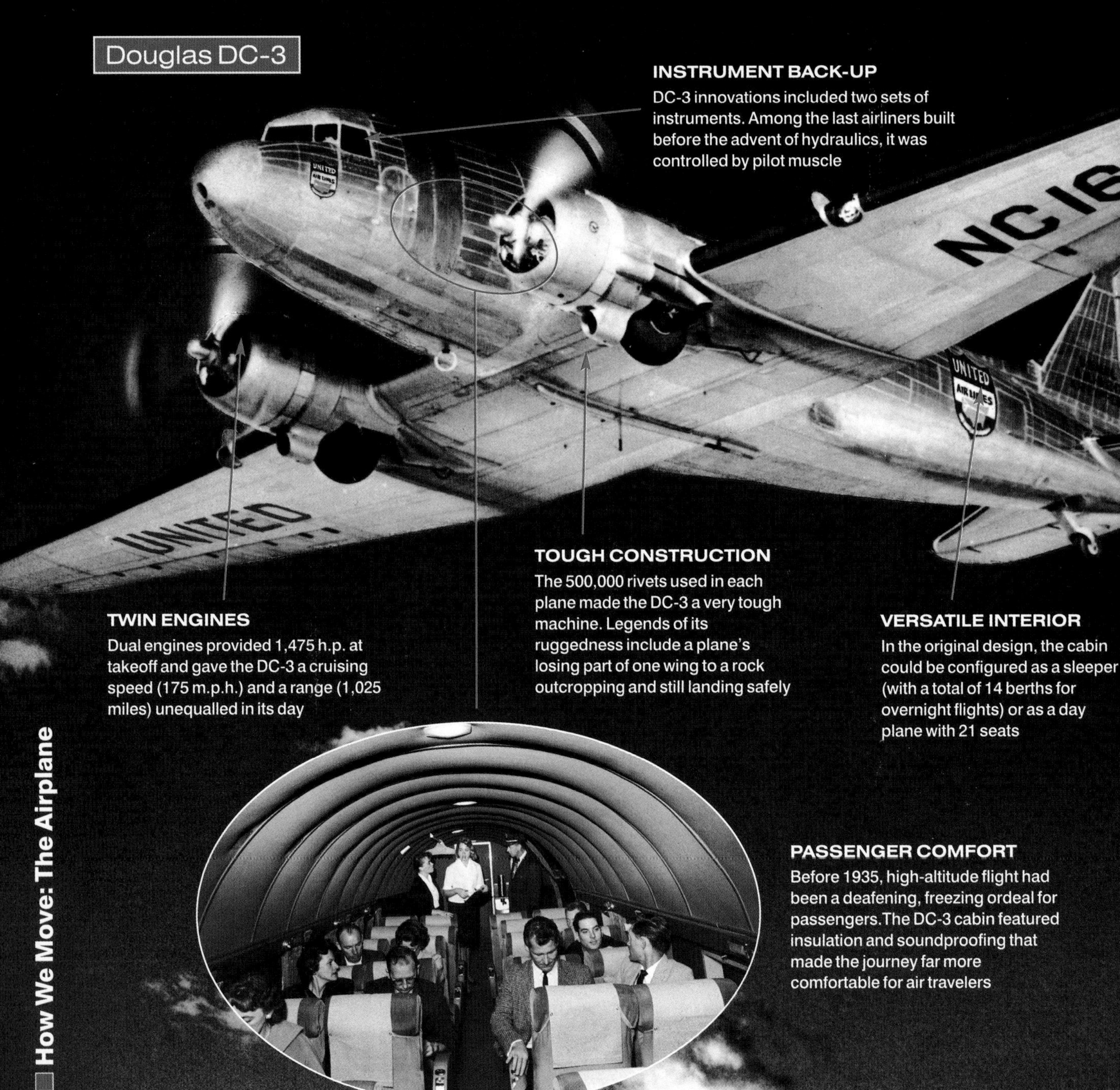

INSTRUMENT BACK-UP
DC-3 innovations included two sets of instruments. Among the last airliners built before the advent of hydraulics, it was controlled by pilot muscle

TWIN ENGINES
Dual engines provided 1,475 h.p. at takeoff and gave the DC-3 a cruising speed (175 m.p.h.) and a range (1,025 miles) unequalled in its day

TOUGH CONSTRUCTION
The 500,000 rivets used in each plane made the DC-3 a very tough machine. Legends of its ruggedness include a plane's losing part of one wing to a rock outcropping and still landing safely

VERSATILE INTERIOR
In the original design, the cabin could be configured as a sleeper (with a total of 14 berths for overnight flights) or as a day plane with 21 seats

PASSENGER COMFORT
Before 1935, high-altitude flight had been a deafening, freezing ordeal for passengers. The DC-3 cabin featured insulation and soundproofing that made the journey far more comfortable for air travelers

Evolution of the Jet

Jet engines use fans to suck in air, then compress it in a confined chamber, where fuel is added and ignited. This creates high pressure, which seeks to escape. When the air—now under very high pressure and moving at great speed—is vented out the back end of the engine, forward thrust is created.

1939: HEINKEL

Germans Ernst Heinkel and Hans von Ohain flew this jet plane in 1939, but their work was ignored. In 1944 a Messerschmitt 262 became the first jet to fly into combat

CULVER PICTURES

1944: SHOOTING STAR

The first U.S. military aircraft ever to fly faster than 500 m.p.h., the Shooting Star was the first jet bought by the U.S. military in large numbers. On Nov. 15, 1950, a Shooting Star shot down a Soviet MiG-15 over Korea in history's first all-jet dogfight

1952: DEHAVILLAND COMET

The first passenger jet, the Comet suffered a series of 20 crashes, killing 110 passengers, before it was grounded in 1954. The verdict: repeated expansion and contraction of its pressurized cabin resulted in metal fatigue

SEATTLE POST INTELLIGENCER COLLECTION—MUSEUM OF HISTORY & INDUSTRY—CORBIS

1954: BOEING 707

The DC-3 of the jet age, the 707 flew higher (up to 40,000 ft.), faster (more than 600 m.p.h.) and longer (more than 6,000 miles without refueling), with more passengers (as many as 181), than any airliner before it. Left, engines on the assembly line

CULVER PICTURES

Evolution of the Jet

Supersonic transports. Radar-defying fighters. Jets that take off and land vertically. Aircraft innovation, once open to tinkerers in Ohio bicycle shops, is now a billion-dollar industry pursued exclusively by the military and an ever shrinking number of commercial airline builders.

TIMEPIX

TOPHAM—PHOTRI—IMAGE WORKS

1969: HARRIER

Britain's Harrier "Jump Jet" became the world's first practical VTOL (Vertical Take-Off and Landing) craft. It was so admired that it was purchased in large numbers by the U.S. military, which made it part of the Marines' air arsenal

FRANÇOIS BAILE—NICE MATIN—IMAPRESS—IMAGE WORKS

INSIDE THE A380

Scheduled to enter service in 2006, the A380 from European consortium Airbus Industrie, will up the ante for jumbo passenger jets

INTERIOR SPACE
Scaled for larger bodies, seats and stairways provide more wiggle room

ECONOMY CLASS
437 seats

MD-11F Length 200 ft. 11 in. Seating 285 Wingspan 169 ft. 6 in.

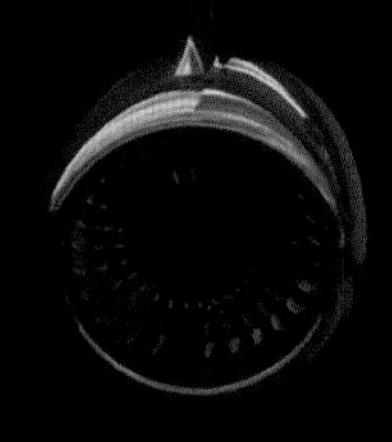

Boeing 747-400 Length 231 ft. 10 in. Seating 420 Wingspan 211 ft. 5 in.

Airbus A380 Length 239 ft. 6 in. Seating 555 Wingspan 261 ft. 10 in.

Productivity of the new Airbus is expected to surpass that of other planes, carrying more payload a greater distance

1960S: V/STOL

This German V/STOL (Vertical or Short Take-Off and Landing) prototype took off on very short runways and landed like a helicopter. The design used one set of engines for take-off and another for cruising. It was unstable, and the nacelles (tilting engines on the wing) were abandoned

TOPHAM—PHOTRI—IMAGE WORKS

1969: CONCORDE SST

Britain and France collaborated on the first supersonic transport. But the Concorde was never profitable, and the plane was grounded for good in 2003, three years after a crash in Paris claimed 113 lives

1982: F-117 STEALTH FIGHTER

The F-117 Nighthawk isn't particularly fast and carries no defensive armaments. It doesn't have to: its armor consists of its invisibility to enemy tracking systems. Among its innovations: a faceted shape that deflects radar, a still-classified radar-absorbing skin, and a shielded exhaust system that leaves heat-seeking missiles with no telltale trail

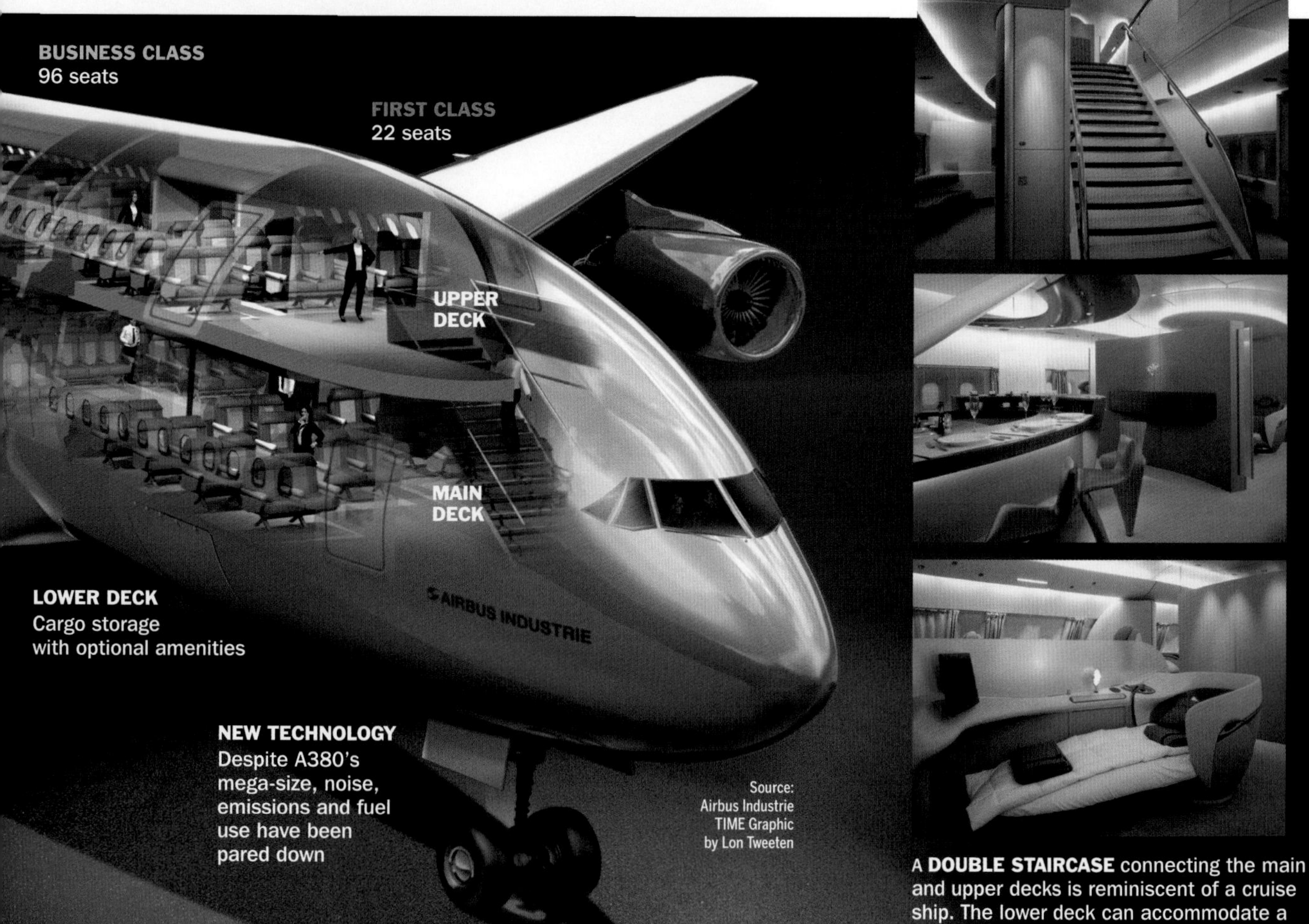

Source: Airbus Industrie
TIME Graphic by Lon Tweeten

A **DOUBLE STAIRCASE** connecting the main and upper decks is reminiscent of a cruise ship. The lower deck can accommodate a variety of amenities, including **BARS**, casinos and **SLEEPING BERTHS**

PICTURES INC.—TIMEPIX

FERDINAND VON ZEPPELIN
A retired Prussian army officer (who served with the Union Army during the American Civil War), he died 20 years before his *Hindenberg,* sister ship to the *Graf Zeppelin,* below, exploded in 1937

Up, Up and Away!

Why should airplanes have all the fun? Blimps, dirigibles and helicopters offer an alternate path to the skies

GRAF ZEPPELIN
Spanking new, the big ship is moved out of its enormous hangar. Below, aviation pioneer Santos Dumont circles the Eiffel Tower in an early airship race, 1901

TIMEPIX

For years after the Wright brothers' first flight, the public was amused—but little more—by the new flying machines. After all, the world already boasted a fleet of passenger aircraft. Faster than steamships and superior to airplanes for safety and comfort, lighter-than-air ships promised to challenge railroads as the future of long-distance travel.

In 1852 French inventor Henri Giffard built the first powered lighter-than-air craft, joining a steam engine and propeller to a gas-filled balloon similar to those flown by the pioneering Montgolfier brothers a century before. Giffard's idea was perfected in 1900, when Germany's Count Ferdinand von Zeppelin attached a rigid frame to an airship, creating the first dirigible, which carried heavier loads and was much easier to control.

By the mid-1920s, crossing the Atlantic by dirigible was commonplace: the trip took three days, rather than the week's time that even the fastest ocean liners required. In 1929 the *Graf Zeppelin* made history's first passenger flight around the world by air, with 61 people (20 passengers and 41 crew) aboard.

Then, on May 6, 1937, the future went up in flames at Lakehurst, N.J. The culprit may have been the flammable paint on the outer skin of the *Hindenberg* rather than the flammable gas inside. Either way, the tragedy ended the airship's passenger days. Today the old roles of the airship and the plane are reversed: airplanes are serious business, lighter-than-air craft a charming curiosity. ■

Evolution of the Helicopter

In the decades after fixed-wing planes first appeared, aviation pioneers began experimenting with rotor-driven aircraft that could take off and land vertically, hover in mid-air, even fly backward. These helicopter pioneers drew their inspiration from the seeds of the maple tree, which rotate as they fall.

BETTMANN CORBIS

1923: AUTOGYRO

Spain's Don Juan de la Cierva theorized that rotating blades mounted atop a fixed-wing aircraft wouldn't need power from the engine—the flow of air through the blades would create lift. His autogyro flew, but couldn't move fast or haul much weight: a dead end

BETTMANN CORBIS

1923: MULTIROTOR

France's Etiénne Oehmichen hoped to give his early chopper added lift by using balloons and multiple rotors. He set early distance records, but the design proved unstable

MANSELL COLLECTION—TIMEPIX

BETTMANN CORBIS

1940: SIKORSKY VS-300

The first practical helicopter was a German design that flew in 1936, but Russian-American Igor Sikorsky added the small tail rotor and perfected the balance systems still used in modern helicopters

GRAHAM MEGGITT—ROYAL NAVY—GETTY

UH-60 BLACK HAWK

Built by Sikorsky, the Black Hawk has been the U.S. Army's primary chopper since 1978. Top speed: 163 m.p.h. Crew: three: Cargo: 11 fully equipped troops. At right, hauling British Royal Marines in Afghanistan, 2002

They Waived the Rules

A trio of iconoclasts creates ships powered by turbines, atoms and air

At sea, utility trumps beauty. The form of each specialized craft—bulky oil supertanker, sleek sub or top-heavy hovercraft—recapitulates its function.

■ THE TURBINE-POWERED SHIP

The most impressive guest at the naval review marking Queen Victoria's Diamond Jubilee on June 26, 1897, was uninvited. As the British fleet paraded before Her Majesty, a smaller, steel-hulled vessel appeared on the horizon, belching flames from its funnels and speeding directly for the armada.

The *Turbinia*, engineer Charles Parsons' revolutionary craft, darted into the stately cavalcade of 150 ships, then back out. Several times it wove between the fastest ships in the world, then literally ran rings around them. The *Turbinia* clocked a record-breaking 34.5 knots that day, faster than any other ship then afloat could sail.

The secret: Parsons' revolutionary new engine. Like the ships it shamed, the *Turbinia* was powered by steam, but she had no pistons moving up and down. Instead, her engines applied the pressure of steam directly to a motor equipped with turning fans, a turbine, which then turned the ship's screws.

Parson's breakthrough design remade life on shore as well as at sea. Today, no matter what fuel is used as a power source—oil, coal, natural gas or nuclear—most of the world's electricity is generated by steam turbines.

■ THE NUCLEAR SUBMARINE

Less than a year after the atom bomb hastened the end of World War II, Captain Hyman Rickover appeared at one of the Manhattan Project's secret research facilities and introduced himself with the words, "I am Rickover. I am stupid."

Rickover, who had spent the war designing technologically advanced ships, was anything but. He was already dreaming about what would become his life's work: freeing ships from the logistical tyranny of having to break off operations

TON ARCHIVE—GETTY

AIRBORNE

A one person hovercraft circa 2003. The basic principles—lift and propulsion—haven't changed since 1955. Hovercrafts are energy-efficient because they move without friction

Steering fins

Large fan provides forward thrust and also forces air under the hovercraft

Motor

Inflatable skirt holds air under hovercraft.

Air under hovercraft is at a higher pressure than the atmosphere, lifting the craft off the surface

TIME Graphic by Joe Lertola

1955: HOVERCRAFT

Cockerell's first model of the hovercraft is above. Top left, an early demonstration of what the inventor's original patent described as "neither an aeroplane, nor a boat, nor a wheeled land vehicle"

©SCIENCE MUSEUM—TOPHAM—THE IMAGE WORKS

HULTON ARCHIVE—GETTY

1894: *TURBINIA*
Before unveiling his superfast ship to the world, designer Parsons spent three years studying the problem of propeller cavitation—erosion of the screws caused by the high speeds that no ship had ever attained before

periodically to replenish their fuel in port. Assigned to civilian status, he led a team of engineers in developing a prototype land-based reactor that could be adapted for use at sea.

Four years later, with his reactor up and running, Rickover went back to uniformed duty to implement the second part of his plan—convincing the U.S. Navy to build a submarine that could house his new power plant. After three more years, the *Nautilus* (after Jules Verne's imaginary craft) was ready for sea trials. It set sail from Groton, Conn., in January 1954 and traveled more than 62,000 miles on its first load of atomic fuel.

Today more than 200 U.S. Navy vessels are reactor powered.

■ THE HOVERCRAFT

After helping invent radar, Briton Christopher Cockerell grew interested in making watercraft move faster. The key, he believed, was to separate the water from the craft, reducing friction between the surface of the sea and the hull, or bottom, of the boat.

But how? Wondering if a cushion of air could possibly heft a boat, Cockerell placed one tin can inside another, then switched a vacuum cleaner into reverse and directed its airflow into the space between them. Liftoff! The inner can levitated.

After several years of tinkering, Cockerell built a working model, a vehicle shaped like an inverted bowl with a high-powered fan mounted on top. The fan drew air in and forced it downward, causing the ship to levitate; air thrust aft drove the ship forward. Cockerell unveiled his craft in 1959. Intrigued by its military potential, the British government took control of the project and shoved Cockerell aside, paying him virtually nothing for his breakthrough work. ■

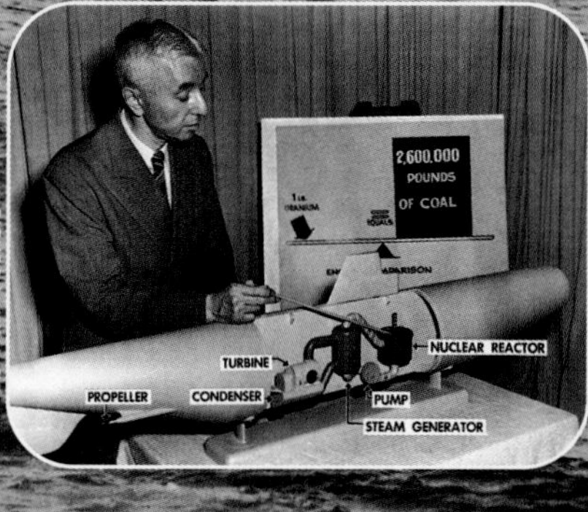

BETTMANN/CORBIS

1954: NUCLEAR SUBMARINE
The *Nautilus* is launched from the Groton, Conn., shipyard of Navy contractor the Electric Boat Co., left. Two years earlier, Rickover, above, explained how a reactor would power the revolutionary vessel

BROWN BROTHERS

Taming the Iron Horse

The railroad could power the Industrial Revolution—if it could be controlled

He called it the "Rocket." Sure enough: in 1830, when Briton George Stephenson's coal-fired locomotive worked up a head of steam and reached speeds of 30 m.p.h. even as it bore several tons of freight and passengers over parallel rails, the Industrial Revolution achieved liftoff.

The railroad was a brilliant three-way hookup of the wheel, the steam engine and the iron rail: motion, power, direction. Early on, train trouble also came in threes: frequent crashes on overcrowded lines; the inability to stop a fast-moving train quickly; and the lack of a reliable, safe method of coupling and uncoupling cars.

In 1865 engineer Ashbel Welch solved the first dilemma, linking his blocking system to traffic signals along the track. If a trainman missed a stop signal, a lever on the blocking system would automatically rise up and engage the braking system.

Still, in a machine age, the braking system was manual: a brakeman hauling on a line in every car! In 1869 George Westinghouse introduced air brakes, operated by a single switch in the locomotive. A year before, store clerk Eli Janney patented his "knuckle" coupler, a new way of automatically connecting and disconnecting rail cars, based on the motion of a left and right hand hooking fingers together.

In the eyes of the public, these essentials (improved, they are still in use) made rail travel safe. George Pullman made it civilized, reimagining the "iron horse" as an iron home with his sleeper car. For the next 80 years, railroads reigned as the world's dominant mover of both freight and folks. ■

CULVER PICTURES

BROWN BROTHERS

GEORGE PULLMAN
A former cabinet maker and shopkeeper, Pullman popularized long-distance rail travel with his innovative sleeper and dining cars and elegant salon cars, like the one at left. But his hard-nosed labor policies led to a major strike in 1894

DALLAS AND JOHN HEATON—CORBIS

Evolution of the Trolley

Railroads, initially long-distance movers, were soon adapted as mass transit for cities. By 1916, streetcars ran along major boulevards in the world's largest cities. Four decades later, most of these street systems were gone, replaced by buses. But railroads (disguised as subways) are still the transit spine of many a city.

HULTON DEUTSCH COLLECTION—CORBIS

1883: ELEVATED ELECTRIC TRAIN

Two years after the world's first electric railway began running in Berlin, an elevated electric train, designed by British inventor and electricity pioneer Magnus Volk, began running along the seashore in the British resort town of Brighton. It remains in service, 120 years later

BROWN BROTHERS

ELECTRIC TROLLEY

Horse-drawn streetcars gave way to the electric trolley technology invented in 1888 by U.S. motor wizard Frank Sprague, in which a metallic arm reaches up to an overhead line that supplies power to the motor. Other versions put the power into a special electrified track. Above, riders on the Coney Island line, 1910

1873: CABLE CAR

British immigrant Andrew Hallidie invented cable-drawn transit after watching the horses that pulled wagons up San Francisco's steeply graded streets drop dead from exertion

BROWN BROTHERS

1964: BULLET TRAIN

Japanese engineers built the first high-speed railroads, the electric-powered "bullet trains," or *shinkansen.* The nose was modeled after a DC-8 airplane. Newer models, like this one passing Mount Fuji in the 1980s, can exceed 200 m.p.h.

Glide Path

Walking is one of the most basic human activities—but who says it couldn't use a little enhancement?

A REALLY BIG SHOW
Otis shows off his invention, with a touch of Barnum. His "improvement in hoisting apparatus" made tall buildings feasible, but skyscrapers did not become common until building technology advanced in the late 19th century. Today, elevators move the equivalent of the world's entire population every 72 hours

Early in the 1850s, Elisha Graves Otis was working in a New York City bed factory whose owner needed a safe way to get men and merchandise from one floor to another. Elevators had existed, in one form or another, since around 300 B.C., but they were considered deathtraps because of their propensity for plunging down their shafts at (literally) breakneck speeds.

Otis, a tinkerer, rigged up a used wagon spring and laid it horizontally across the top of an elevator cab, with heavy metal latches on either end. A hinged brace, attached to the rope that held up the elevator, compressed the spring. If the rope snapped, the brace would open and the spring would expand, driving the latches into the vertical guide rails on the sides of the shaft.

Otis' device worked, but the bed factory didn't: it went bankrupt. So Otis decided to join the California Gold Rush. He was delayed by two other New York City factory owners, who had heard of his contraption and wanted their own "safety elevators." Foreseeing better prospects in elevator shafts than mine shafts, Otis stayed put and formed the Otis Elevator Co.

Business trickled in until the Crystal Palace Exposition of 1854, where promoter P.T. Barnum paid Otis $100 to demonstrate his invention. As

OTIS ELEVATOR COMPANY. INSET: HULTON ARCHIVE—GETTY

The Escalator

Jesse Reno invented the first primitive escalator in 1891, as a ride for the Coney Island amusement park. It was called the Reno Inclined Elevator. The first modern moving staircase, seen under construction here in 1899, was developed by inventor Charles Seeberger—who also coined the term escalator—for the Paris Exhibition of 1900. Both Reno and Seeberger sold their patents to the Otis company, which retained a trademark on the word escalator until the 1950s

shown in the contemporary drawing at left, just as Otis ascended to the top of the cavernous hall in an elevator, an axe-wielding assistant cut the rope that was holding it aloft. The crowd shrieked as the car dropped, but Otis—caught in mid-fall—shouted, "All safe, gentlemen; all safe!"

Three businessmen in the crowd placed orders on the spot; 22 more commissions came in over the next two years. But Otis did not live to see his invention become ubiquitous: he died in 1861, at 49. It was left to his son Charles to invent (in 1878) the speed-control mechanism that made elevators safer still by slowing them gradually in an emergency, rather than triggering a sudden stop. The development by Otis engineers of the automated controls that made elevator operators obsolete would await the 1950s, when solid-state circuitry and the first rudimentary computers bred change. ■

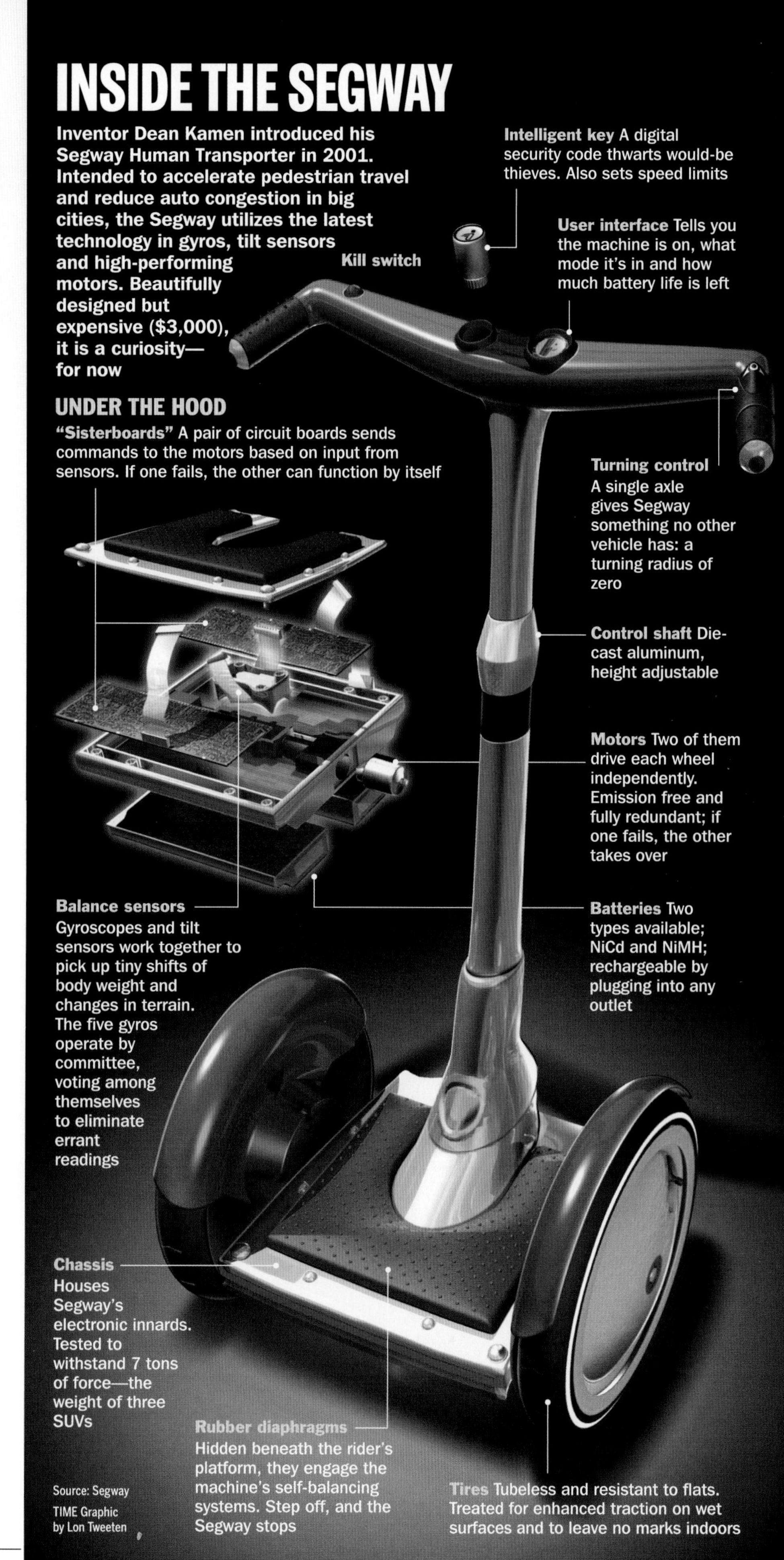

EASY RIDER

The Segway's code name was "Ginger" (for Ginger Rogers) at DEKA Research. But the two-wheeler dances courtesy of ten microprocessors cranking out three PC's worth of juice

Profile

Tilting with Windmills

Call him the anti-Ford: Dean Kamen, a genius with a conscience, is off and rolling on his plan to make the automobile the wave of the past

HE RODE IN ON A SEGWAY, EXPLODING INTO THE public imagination late in 2001, leaning and tilting and buzzing around on the new personal transportation device he hopes will revolutionize human movement. Dean Kamen hadn't attracted much public notice before he suddenly became very famous, but this 52-year-old college dropout has been inventing breakout technologies since he was a high school student on suburban Long Island, N.Y., in the mid-1960s.

Kamen had a summer job at a small factory. "We were making components for professional audiovisual equipment," he recalls. "I had an idea for a control box that could manage sound and light shows. By putting together a few boards and a few switches, I could build something that would handle the work that had been done by a dozen grown men pulling huge levers backstage."

So Kamen followed Henry Ford, Thomas Edison and countless others whose ideas remade the world. "I quit my job," he recalls, "and went off to build one of these things on my own."

First box in hand, the youngster showed up unannounced at New York City's Hayden Planetarium and offered to demonstrate it. "They couldn't believe it," Kamen remembers of the staff's reaction. Within weeks, the planetarium had purchased, for thousands of dollars, a contraption Kamen had assembled from simple parts costing a few hundred.

Before the end of the year, Kamen had orders for more than $10,000 worth of his boxes—not bad for a 16-year-old. Most teenage boys armed with that kind of money could think of several ways to get into trouble. What Kamen wanted to get into was "a really nice oscilloscope, some high-end meters and a few pieces of good calibration equipment." Well, there was an agenda. "I used some of the money to buy the equipment I wanted," he says, "and some to send my parents away on vacation. The problem was, I needed a place to put all this new equipment. So while my parents were away, I hired contractors to come jack the house up while they expanded the basement into a full workshop." Kamen's father was "not amused" when he came home and found the rear façade of his house perched on timber stilts.

Next stop: medicine. "My brother was in medical school," he says. "And he was complaining that there was no device for administering medication to children automatically, as there is for adult patients." Within two years—while he was struggling to stay interested in his classwork at Worcester Polytechnic Institute in Massachusetts—Kamen had perfected his AutoSyringe. For children with cancer, who can require many injections a day, it was a life-changing innovation. Kamen left college: a few years later, he sold AutoSyringe to a large health-supply firm for several million dollars.

The next product from the Kamen workshop (founded in 1982, DEKA Research now employs 200 engineers): a machine to provide mobile dialysis treatment, freeing kidney patients from clinic visits. In 1999 DEKA introduced a wheelchair, the IBOT, that can climb stairs. Kamen's work with gyroscopes on the IBOT (he calls it a wearable robot) inspired the research that led to the Segway, which is both product and crusade for its inventor. He told TIME in 2001: "It makes no sense at all for people in cities to use a 4,000-lb. piece of metal to haul their 150-lb. asses around town."

"It makes no sense at all for people in cities to use a 4,000-lb. piece of metal to haul their 150-lb. asses around town."

Along the way, Kamen bought a private island off Connecticut (with lighthouse), a complex of 19th century mill buildings in New Hampshire where he and his colleagues cogitate and tinker, and two helicopters, his second-favorite personal transportation device.

What's next? Kamen hints mysteriously at his team's work on a new type of efficient, nonpolluting engine: after years of dead ends, success seems closer. "Before I take something on," the inventor says, "I want to be sure it will benefit a lot of people and probably won't get done unless we get involved. If you look at the competition to build the next generation of microchips, there are a lot of smart people working on that problem. It's going to get done sooner or later. So it doesn't matter who wins, because the public will still have the answer."

"Life is short," the Quixote of the Segway insists. "I want to work on things that matter." ■

GREAT INVENTIONS

How We Communicate

Bell Telephone, patent drawing, 1876

3

1881: THE SWITCHBOARD
Hand-connected switchboards, like this pyramid design in Richmond, Va., were used to relay burgeoning telephone traffic until Almon Strowger invented the Automatic Telephone Exchange in 1891. But his invention, which eliminated the need for an operator on local calls, didn't catch on until the 1920s

The First World Wide Web

Turning the human voice into electricity and sending it down a wire, the telephone abolished distance, united those far apart and helped define the modern era

UNDERWOOD & UNDERWOOD—CORBIS

The world changed on May 24, 1844, when Samuel F.B. Morse instantly transmitted the words "What hath God wrought?" from the U.S. Capitol to Baltimore, using only a series of long and short bursts of electric current (the dashes and dots of his namesake code) carried over a wire. For long after, engineers and inventors around the world were haunted by a nagging question: If messages could by sent over a wire by electrical impulses, why not a human voice?

By 1875, Alexander Graham Bell, a Scots immigrant to the U.S. and speech teacher, had been working for three years on what he called a "harmonic telegraph," which would send as many as 30 or 40 messages over a single wire simultaneously, by varying the electrical current used for each message. Because the voltage would oscillate, moving up and down in waves, rather than shutting on and off (as telegraphs did), the transmissions began to resemble actual sounds, instead of the dots and dashes of a telegram. "If I can get a mechanism which will make a current of electricity vary in its intensity as the air varies in density when a sound is passing through it," Bell said to his assistant Thomas Watson in the spring of that year, "I can telegraph any sound, even the sound of speech."

J-L CHARMET—SPL—PHOTO RESEARCHERS

Although the idea had come to him suddenly, making it work proved maddeningly difficult. Bell and Watson stumbled upon momentary success for the first time on June 2, 1875, when Bell briefly heard through his prototype the sound of Watson plucking on a spring in the next room. But they could not duplicate the effect consistently. Bell filed for a patent on Feb. 14, 1876, even though he hadn't yet produced a working phone, only hours before rival Elisha Gray filed his own request. Shortly after he applied for his patent, Bell made the crucial discovery that a

THREE BELL PHONES

- Top, Bell's first working phone, over which he said, "Watson, come here, I wish to see you," on March 10, 1876
- Center, an 1875 "gallows" phone, first to use a diaphragm (instead of a spring) to pick up acoustic vibrations and transmit them as electric current. It barely worked
- Right, a more familiar 1877 design was used in a demonstration for Queen Victoria

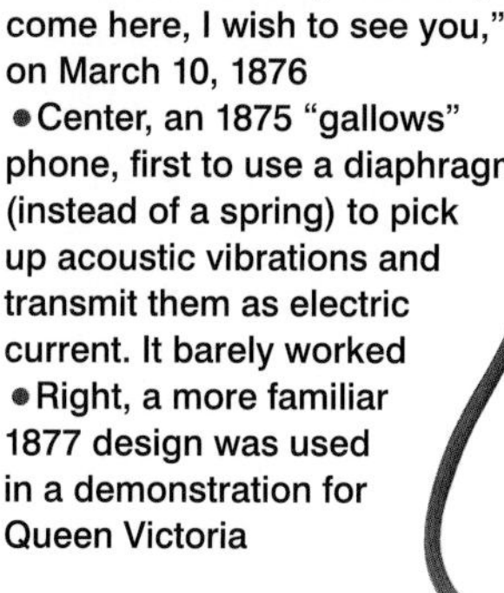

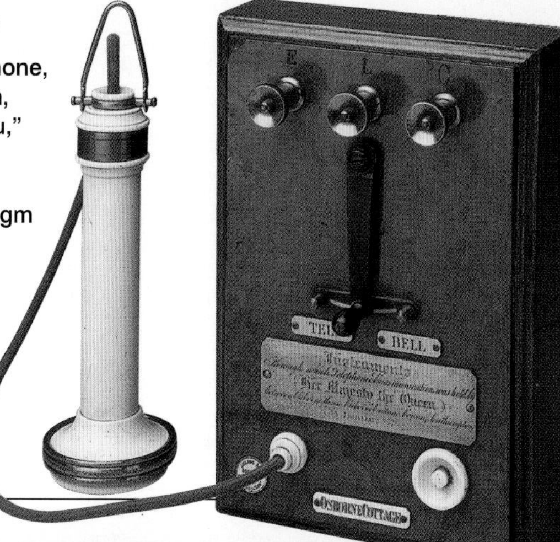

©SCIENCE MUSEUM—TOPHAM—THE IMAGE WORKS

Down to the Wire

CULVER PICTURES

ALEXANDER GRAHAM BELL

U.S. law changed in 1870, allowing for the initial filing of a patent application even if the inventor didn't yet have a working model. All that was needed was a set of detailed drawings—and that's all Bell had in February 1876. He and Watson didn't succeed in making a working telephone until more than a month after the patent was filed and several days after it was granted

SPL—PHOTO RESEARCHERS

ELISHA GRAY

On the same day Bell filed for his patent, rival Elisha Gray filed a "caveat"—a kind of pre-patent that protects an idea for a limited time. But Gray's filing followed Bell's by several hours. Gray went on to co-found Western Electric (now Lucent) and invent an early version of the fax machine. As one of his partners said, "Of all the men who didn't invent the telephone, Gray was the nearest."

1889: THE PAY PHONE
William Gray invented the pay phone after his wife went into labor and he needed to call an ambulance, but the only nearby business with a phone denied him access. Below, phone booths in New York City, 1923

BROWN BROTHERS

wire partially immersed in a conducting liquid, like battery acid, could be made to vary its electrical resistance and produce an undulating current. If the current was triggered by a diaphragm vibrating in response to sound, those undulations would vary in proportion to the sound waves, duplicating human speech. Although the familiar tale of Bell's calling for Watson after spilling battery acid on his hand is probably apocryphal, Watson did hear Bell's voice over a wire for the first time on March 10, 1876.

The debate over whose work led to the creation of the telephone would play out for decades. In the end, the credit went to Bell, who (along with his partners in the Bell Co.) converted the new device into a business. With a notable assist from Thomas Edison, whose 1877 patent for a transmitter diaphragm made the telephone truly practical, Bell's gadget soon replaced the telegraph it was originally intended to augment. By 1878, local switching stations had been established in most major U.S. cities. By late 1879, the Bell Co. had connected 133,000 telephones. But as the lines mushroomed, asking the operator for another customer by name became impractical. So in 1879, Bell issued the first phone numbers.

Customers couldn't dial those numbers themselves until 1891, when Almon B. Strowger, a Kansas City undertaker—convinced that party-line operators were routing his incoming calls to competitors—devised an automatic switch that eliminated the need for an operator on most local calls. Even so, most lines weren't automated until the 1920s. Phone engineers were fighting a larger and longer problem. Because electrical impulses (like sound waves) fade over distance, telephone users around the country could talk to those with phones in their own city but not in another.

The long-distance age began in 1899, when George Campbell, a Bell Co. scientist, invented loading coils, which increased a signal's distance before it began to fade: New Yorkers could now call Denver. In 1915, engineer Harold Arnold invented the first practical electrical amplifiers, allowing calls to reach anywhere phone wires did. New Yorkers could now chat with San Franciscans, albeit at the lofty rate of $20.70 for the first three minutes. And they say talk is cheap. ■

WIDE WORLD PHOTOS

1905: ROTARY PHONE

Almon Strowger's dial phone, right, built to work with his Automatic Telephone Exchange, was replaced by French inventor Antoine Barnay's version, which placed the lower numbers on top and matched each numeral with three letters of the alphabet

SCIENCE MUSEUM—TOPHAM—THE IMAGE WORKS

1950: ANSWERING MACHINE

A model shows off the "automatic electronic secretary," above. It lifted the receiver, played a greeting off a vinyl record, then recorded the message in magnetic form. The Ansa Phone, invented by Japan's Kazuo Hashimoto and introduced to the U.S. in 1960, was the first system to intercept and tape a call without lifting the receiver

1972: PORTABLE PHONE

Beam me up, Scotty! A model shows off a "portable radio-telephone" made by Pye Telecommunications. Such early "portable" phones were bulky—and limited by the short range of their radio transmitters

MOTOROLA

Evolution of the Cellular Phone

In 1947 Bell Labs scientists realized that the basic problem of radio-based telephony—range—could best be solved not by boosting the power of transmission towers and portable phones (the initial approach, under which just a few towers would cover an entire city) but rather by increasing the number of those towers, each responsible for a specific "cell" on a geographic grid, and working at a lower power. The technology languished until the late 1960s, when the Federal Communications Commission first set aside sufficient radio bandwidth to make cellular systems practical. The race to design the first working cell phone was effectively won in the early 1970s by Motorola's Martin Cooper (who placed the first call to a rival at Bell Labs), although trials did not end until 1982, when the U.S. government authorized nationwide cellular service.

THEN: 1983		NOW: 2003
Dynatac	**MODEL**	Samsung A500
3 lbs.	**WEIGHT**	3.6 oz.
$3,000	**PRICE**	$299
None	**VISUALS**	Color LCD display
"Inter what?"	**INTERNET**	High-speed connection

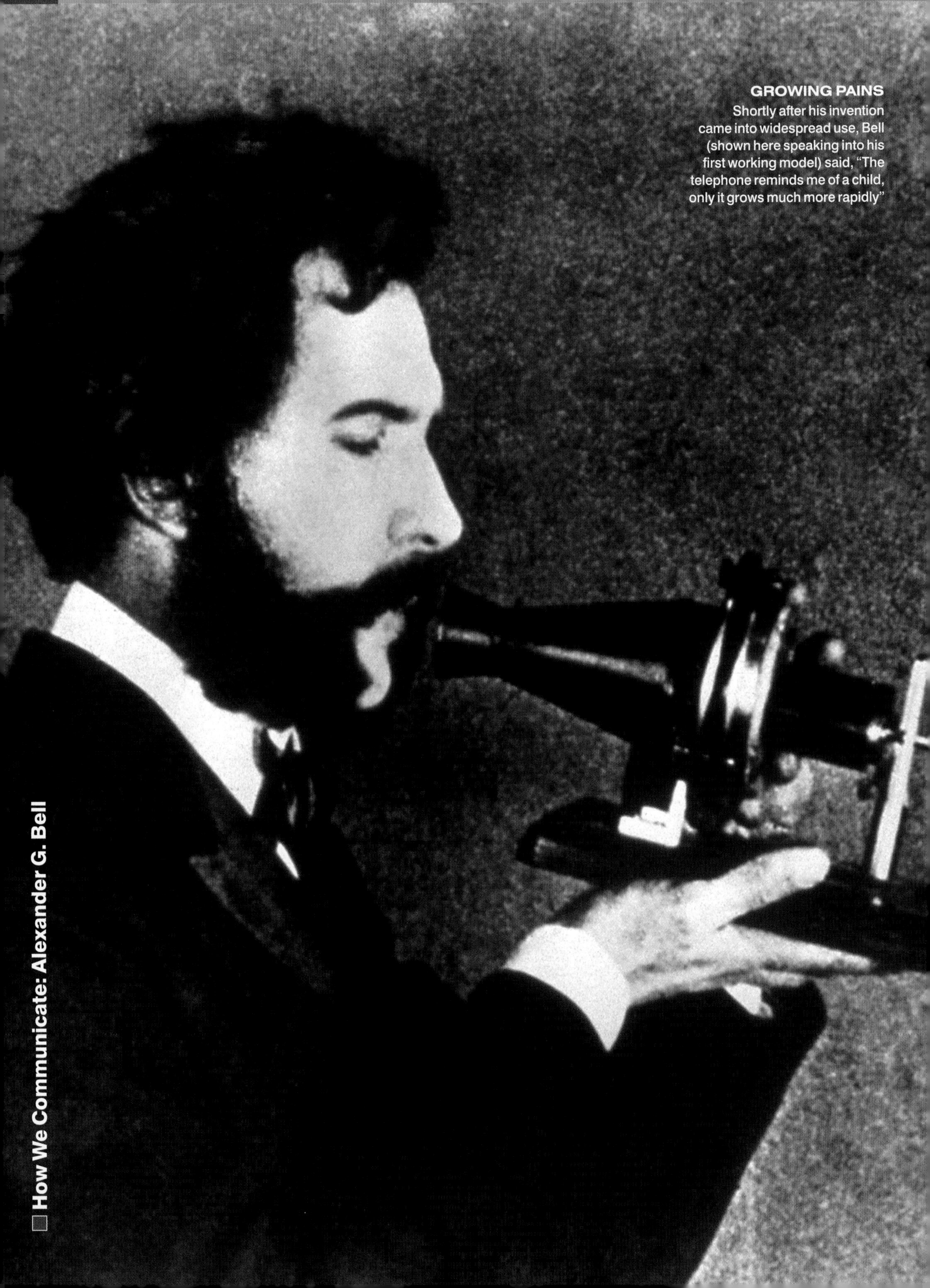

GROWING PAINS
Shortly after his invention came into widespread use, Bell (shown here speaking into his first working model) said, "The telephone reminds me of a child, only it grows much more rapidly"

Profile

Teacher of the Deaf

A flighty dreamer but a devil for work, Alexander Graham Bell appropriated the secrets of speech—and connected the world

GARDINER HUBBARD, ALEXANDER GRAHAM Bell's future father-in-law, had strong words for the young man in a letter sent in the mid-1870s. "If you could work as other men do," he admonished the teacher of the deaf, "you would accomplish much more than with your present habits. While you are flying from one thing to another, you may accidentally accomplish something, but you probably will never perfect anything."

What Hubbard didn't get, but Bell did, was that flying from one thing to another was precisely the point. In his struggle to invent the telephone (in which Hubbard had invested heavily), Bell searched for clues in places others would never have thought to look. He spent months studying how the phonoautograph, a device crafted out of a human ear harvested from a cadaver, could make a lever jump in response to sound waves: the louder the sound, the stronger the twitch. This inquiry led Bell to envision a device that substituted an artificial membrane for the eardrum and converted sound into electrical energy rather than the lever's mechanical energy. These ideas—variable resistance and the conversion of words into electricity—produced the telephone.

Bell brought to the pursuit of telephony a working knowledge of telegraphs and electricity, as well as the fruits of his "flying about"—a deep background in acoustics and the science of speech. His father Alexander M. Bell was a renowned teacher of the deaf who in 1864 created the Visible Speech system, a universal alphabet of symbols to represent the action of the mouth, tongue and throat in producing sounds.

From the time of his son's birth in 1847, the elder Bell trained Alexander in the ways human anatomy produces and perceives speech. As a teenager, Bell experimented continually with the voice. With his brother, he rigged up a machine that pumped air through a voice box taken from a slaughtered sheep and manipulated it to pronounce "mama." It was an impractical novelty—yet a sign of things to come.

Whatever Bell lacked in blinkered focus, he made up for with a ferocious work ethic. "There are no unsuccessful experiments," he said as he struggled to transmit the human voice. "If we stop, it is we who are unsuccessful, not the experiments."

After the patent for the telephone was secure and the business of wiring America began, Bell did something that Thomas Edison or the Wright brothers would have found unthinkable: he walked away. Leaving the company under Hubbard and another partner, Bell married Mabel Hubbard in 1877 and set sail for what proved to be an extended honeymoon in Europe. Upon his return, Bell decamped to Nova Scotia, where he spent most of the rest of his life working on new inventions, mastering aeronautics and hydrofoil engineering (a boat of his design set a world speed record) and exploring tetrahedrals—lightweight structures that can support immense weight.

"Make me work at anything—it doesn't matter what—so that I may be accomplishing something."

Late in life, Bell wrote in his journal, "Make me work at anything—it doesn't matter what—so that I may be accomplishing something." And work he did, to the very end. In 1922, when death came to the man who, to the end of his life gave "teacher of the deaf" as his profession, all the telephones in the U.S. and Canada went silent for one minute. It was a quiet tribute to Hubbard's dreamer, a man who managed not only to accomplish something but to perfect it, as well. ■

The Wonderful Wireless

Success, it is said, has many fathers. The radio was one hell of a success

Waves. Invisible waves, radiating across what scientists then called the "luminiferous ether." It sounded like science fiction. Yet after 1864, when British physicist James Clerk Maxwell predicted the existence of radio waves as a form of electromagnetic radiation, and theorized that they must move at the speed of light, he sounded the starting pistol for a race that would engage the world's most brilliant engineers and physicists.

Maxwell's theory set telegraph engineers looking for a way to transmit Morse code signals without the aid of all those cumbersome (and very expensive) wires. The race heated up in 1893, when German physicist Heinrich Hertz proved that Maxwell's waves were real. Suddenly, a new device that would earn a vast fortune and a place in history for its creator seemed within reach.

More than any of the other inventions that shaped modern life, radio was the product of diffuse efforts by a large number of geniuses, most of whom never met one another and many of whom engaged in backbiting, sabotage and outright theft of ideas—invention's grubby underbelly.

Conventional wisdom has it that Italy's Guglielmo Marconi invented radio, sending a signal across a room, in 1894. Serbian-American Nikola Tesla, who drew up plans for a working model two years earlier, might disagree. So might France's Edward Branley, who built the first detector of radio waves in 1893. Or Britain's Oliver Lodge, who also sent a verified

HULTON ARCHIVE—GETTY

SARNOFF
Savoring early fame, the boy who would become broadcasting's visionary re-enacts his role in receiving wireless distress signals from the *Titanic,* sinking hundreds of miles away

MARCONI
As a teenager, he used electromagnetic waves to ring a bell on the far side of a room in his mother's home

BETTMANN-CORBIS

message by radio in 1894. Or Russia's Alexander Popov, who demonstrated the first radio receiver in 1895.

By the early 1900s, however, Marconi had consolidated his position as the regent of radio. He had filed or bought key patents and was building a thriving business in sending telegrams without wires. In 1901 he thrilled the world by sending the first radio message (in Morse code) across the Atlantic. Marconi would repeatedly be accused of stealing some of "his" innovations (such as a coil designed by Tesla and a tube that detected radio waves conceived by Indian physicist Jagadis Chunder Bose). Still, it was Marconi, not his rivals, who nabbed a U.S. patent for radio in 1904—and the Nobel Prize in Physics in 1909.

Up to this time, everyone seems to have envisioned radio as little more than an improved telegraph, a better way to send Morse code—everyone except Reginald Fessenden. A Canadian physicist, Fessenden invented a continuous-wave voice transmitter in 1905 and first put it to work (with the help of a "triode" tube, a powerful new amplifying device perfected by American Edwin Armstrong) on Christmas Eve, 1906. This first-ever broadcast of the human voice by radio was received by startled wireless operators aboard ships in the North Atlantic, who had expected to hear Morse code.

Inventor Lee de Forest quickly patented a set of designs for a very similar tube, calling it the Audion.

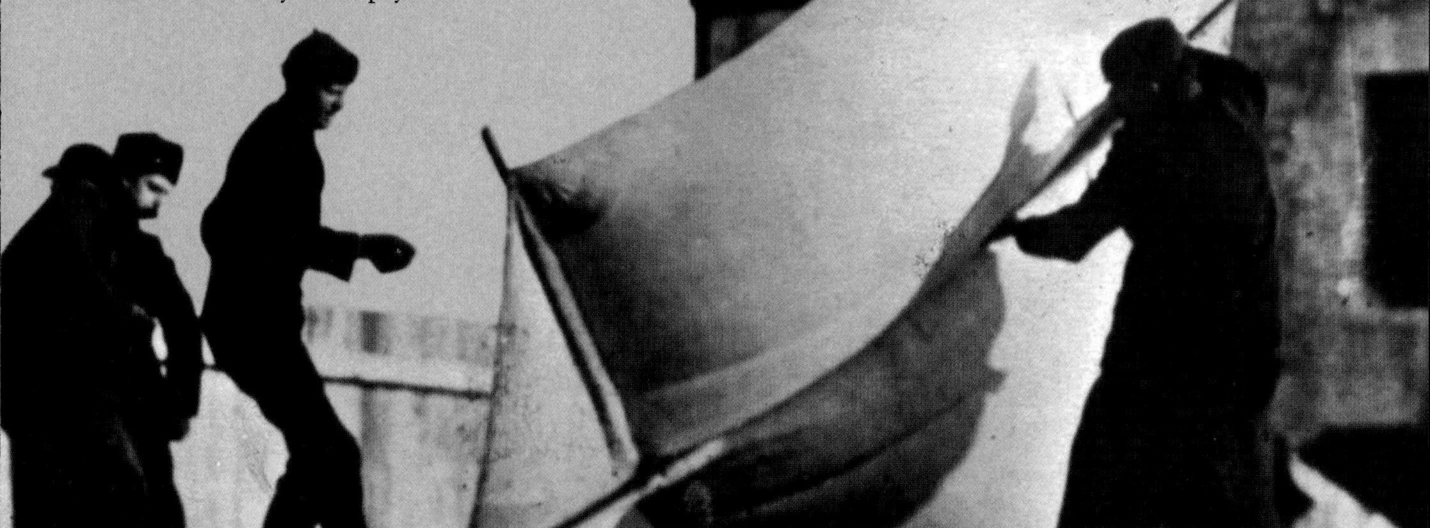

IN BEN FRANKLIN'S FOOTSTEPS
Above, technicians in St. John's, Newfoundland, prepare the kite that will carry aloft an antenna to receive the first transatlantic radio signal. Marconi is visible at the extreme left

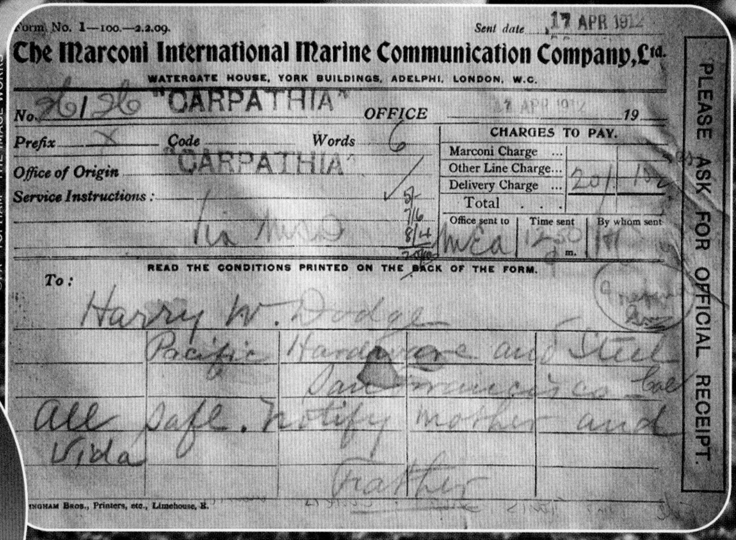

Form No. 1—100.—2.2.09. Sent date 17 APR 1912
The Marconi International Marine Communication Company, Ltd.
WATERGATE HOUSE, YORK BUILDINGS, ADELPHI, LONDON, W.C.
No. 2126 "CARPATHIA" OFFICE 19
Prefix X Code Words 6
Office of Origin CARPATHIA
Service Instructions:
CHARGES TO PAY.
Marconi Charge
Other Line Charge
Delivery Charge
Total
Office sent to | Time sent | By whom sent
READ THE CONDITIONS PRINTED ON THE BACK OF THE FORM.
To: Harry W. Dodge
Pacific Hardware and Steel
San Francisco Cal
All safe. Notify mother and father
Vida
PLEASE ASK FOR OFFICIAL RECEIPT.

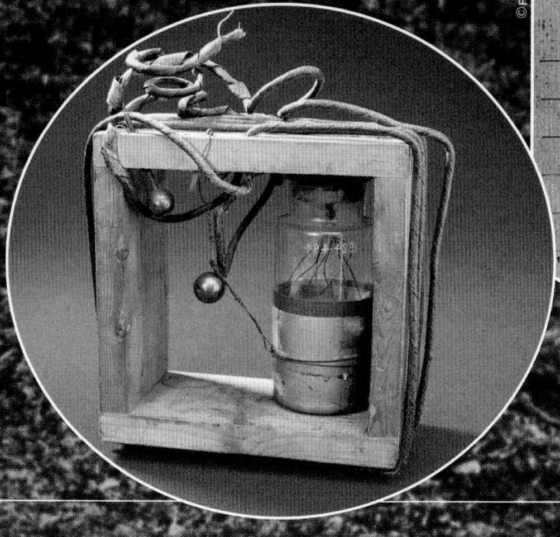

1897: MARCONI RADIO
Left, an early Marconi design, the first that could be tuned to different frequencies. Above, a passenger from the *Titanic* sent this message from the *Carpathia* (which picked up many of the *Titanic's* survivors) via the Marconi wireless station in New York City

BETTMANN CORBIS

The Father of FM

Edwin Howard Armstrong, a New Yorker, was one of radio's founding geniuses. In 1912, at age 22, he invented the regenerative circuit, which amplifies a signal thousands of times, making broadcasting possible. Five years later, while serving in the U.S. Army Signal Corps, he created the superheterodyne, which enhances both receiver tuning and amplification. But the greatest invention of this Columbia University professor was an alternative to the AM (amplitude modulation) format: FM (frequency modulation) radio, which made possible TV, walkie-talkies and cellular phones.

The innovation, which should have been Armstrong's crowning glory, undid him. Because FM is a clearly superior technology to AM radio (modifying wave frequency rather than wave height, it offers better sound fidelity and resists atmospheric static), RCA chief David Sarnoff came to view FM as a threat to his empire, the AM stations that RCA owned and the AM radios it sold.

After encouraging Armstrong and allowing him to place the first FM antenna at an RCA facility atop the Empire State Building, right, Sarnoff worked to suppress the technology. Frustrated by RCA's superior legal and financial muscle, Armstrong committed suicide in 1956. Sarnoff's first words on hearing the news: "I didn't kill Edwin Armstrong."

BROWN BROTHERS

De Forest was later sued by Fessende and Armstrong for stealing their ideas; he lost the case when he couldn't explain to the court how the Audion worked. The tube was that significant: it would become an essential component of the television, and it led to the transistor and the computer.

Once voice transmission over a wide area had been achieved, a different kind of visionary transformed radio from science into commerce. David Sarnoff shot to national fame as a 15-year-old telegraph operator in 1912, when he picked up distress signals from the sinking *Titanic.* The episode inspired Sarnoff, who began hectoring his bosses at the Marconi Co. (later Radio Corp. of America, or RCA) to schedule music and sports programming. This, Sarnoff theorized, would move the

BROWN BROTHERS

1920: FIRST ELECTION BROADCAST
Among the pioneering commercial radio stations in North America, Pittsburgh's KDKA began transmitting from a shack on the roof of a local factory. Some 1,000 listeners tuned in on Nov. 2 to hear that Warren G. Harding had defeated James Cox to win the White House

general public—not just businesses and government agencies—to buy radio sets, which RCA could sell. RCA paid a group of wireless stations to schedule these programs (thus forming the NBC network), and what Sarnoff called a "radio music box" was soon *the* indispensable household appliance. The era of broadcasting had begun.

It was also around this time—just as it no longer seemed to matter—that the argument over radio's paternity was decided. In 1943, the U.S. Supreme Court ruled that Marconi's 1904 patent for radio had been granted in error; his patent was invalidated, and credit for the invention of radio was retroactively awarded to Tesla—who had died eight months earlier, alone and penniless in a seedy New York hotel.

Few noticed and fewer cared about Tesla's posthumous victory: broadcasting companies and a new clutch of inventors were transfixed by an entirely new vision, one that promised to make radio seem like a sideshow. ■

Portable Radio

Radios were exciting enough, you'd think. But no: lovers of the wireless longed to see it unshackled from its power line, so it could be enjoyed on the road, on the run or while tending to agrarian concerns.

BROWN BROTHERS

IN TUNE WITH THE EARTH
Details of the unique predecessor of the portable radio above have been lost; the picture dates to the 1920s, when radios became the rage as the first new mass medium since the advent of motion pictures

Look at the new Motorola Portables!

The handle is a rotating antenna

1955: TRANSISTOR RADIO
Following the lead of Japanese companies like Sony, Motorola was among the first U.S. firms to market portable radios in the '50s

SONY

1979: SONY WALKMAN
Launched as a tape-only machine, it was beefed up in 1982 to include an AM/FM tuner. More than 100 million of them have been sold

BROWN BROTHERS

REMAKING HOME LIFE
Enjoying a broadcast at home in the 1920s. Even at the depth of the Depression, sales of radios held steady; 2 out of 3 homes had one. After World War II, the number of radio stations in America doubled, to more than 2,000, in three years

FARNSWORTH

Above, the ghostly image of Joan Crawford appears inside a cathode ray tube during a 1934 experimental broadcast at the Franklin Institute in Pennsylvania. Farnsworth, inset, conceived his system of electronic television when he was 14 and perfected it when he was 21

BETTMANN CORBIS (X2)

The Image Dissectors

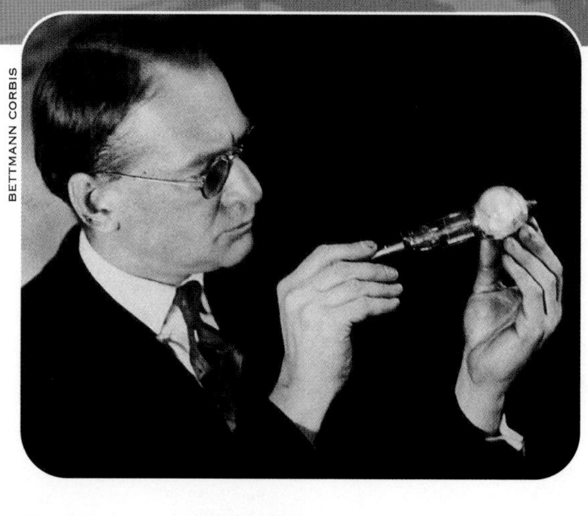

BETTMANN CORBIS

Radio tranformed words into waves. Who would be first to do the same for visions?

It all began in a potato field. In the summer of 1921, a 14-year-old Idaho farm boy was plowing his father's land when he looked back at the straight rows he had just turned up and experienced one of those Eureka! moments that change history. In this case, young Philo T. Farnsworth, who had been dreaming for months about how to add moving pictures to radio, realized that a visual image could be divided into a series of lines and that such lines (precisely arranged areas of light and dark) could be transmitted through the air, just like radio waves.

Since radio and motion pictures had become commonplace a few years earlier, consumers had been clamoring for a combination of the two. But scientists and engineers had been captivated by the idea of sending pictures through the air for far longer, urged on by two important discoveries. In 1872, Joseph May and Willoughby Smith had described the principle of photoconductivity—the idea that the electrical resistance of certain metals varies according to their exposure to light. Then, in 1880, French engineer Maurice LeBlanc theorized that, because the human eye retains whatever it sees for a fraction of a second, a picture transmitted through the air could be broken into parts and sent one piece at a time. As long as those pieces were all put back together in less than a tenth of a second (the duration of what scientists call "visual persistence"), the viewer would see the result as a single image.

But if the idea was not new, Farnsworth's approach was. It took this Mormon farm boy, who had declared at age 6 that he would someday be recalled alongside Thomas Edison and Alexander Graham Bell, to realize that the pieces of LeBlanc's conceptual picture could be horizontal lines, and that such lines could be created by exposing the light and dark areas of a visual image to a photoconductive metal while an electric current ran through it. Farnsworth sketched out the idea for his high school science teacher, Justin Tolman, filling several black-

ZWORYKIN
The Russian American holds his "thermionic photo electric tube" in 1925. When he first saw Farnsworth's Image Dissector, he said, "This is a beautiful instrument. I wish I had invented it myself." RCA later claimed that he had

HULTON ARCHIVE—GETTY

©NMPFT—TOPHAM—THE IMAGE WORKS

BAIRD
The Scottish inventor, pursuing the concept of television based on spinning wheels, managed to transmit the image of puppet faces in 1925. But his work was outpaced by that of Farnsworth and Zworykin

RCA

NBC—GLOBE

1954: COLOR TV

The first practical color television, RCA's CT-100, hit the stores in March 1954, left, two months after NBC inaugurated color broadcasts with the Rose Bowl Parade. Shortly after, a live telecast of *Peter Pan,* above, further promoted NBC's new hue

boards with diagrams and equations. Tolman, impressed, shipped his student off to take university classes after only two years of high school work.

Six years later, Farnsworth's ideas began to take form. He called the crucial part of his invention (the tube within the camera containing the photoconductive plate) an Image Dissector. To display the image, Farnsworth developed a receiver based on the vacuum-tube oscilloscope that German scientist Karl Braun invented in 1897 to display graphs of electrical currents.

On Sept. 7, 1927, Farnsworth and his team aimed their camera at a plate of black glass with a clear line in the center. A light shining behind the glass created a brilliant white stripe that appeared on the screen of the receiver. When the glass was rotated, the image onscreen rotated with it. At last, Farnsworth was able to send to the investors who had bankrolled his efforts for half a decade (and who had badgered him with the incessant query "Does the damn thing work yet?") a telegram containing the words "The damn thing works!"

In a stroke, Farnsworth's "electronic television" demolished decades of labor by proponents of "mechanical television," a system that employed rapidly spinning disks perforated with tiny holes to scan an image, then used similar disks within a receiver to project a dot pattern of light and dark on the screen and re-create it. This trail began with German inventor Paul Nipkow, who created the disk system in 1884, and culminated with Scottish inventor John Logie Baird, who was able to transmit a rudimentary picture in 1925. But his system was doubly hampered: its screen was very small, and its crude picture resolved only a dozen or so lines on the screen, whereas Farnsworth's prototype created far sharper 60-line resolution.

Farnsworth's system also seriously upstaged the work of the one other genius pursuing an electronic approach: Vladimir Zworykin. Working for RCA, Zworykin was granted a patent for his process for electronic television in 1923, even though he hadn't yet made it work. (Farnsworth didn't file his patent until 1927, after he had successfully demonstrated his invention.) But RCA chief David Sarnoff, fresh from hijack-

TOPHAM—THE IMAGE WORKS

1962: SATELLITE BROADCAST

Vice President Lyndon Johnson places a call bounced off Telstar 1, the world's first telecommunications satellite, which handled both TV and phone signals

ZENITH

1954: REMOTE CONTROL
Zenith engineer Robert Adler developed a device that would help viewers ignore ads: the Space Command worked by high-frequency sound waves

ing the development of FM radio from rightful inventor Edwin Armstrong, had no intention of paying Farnsworth for the right to manufacture TV sets.

So RCA claimed that Farnsworth's work infringed on Zworykin's 1923 patent. (In fact, Zworykin did finally create both a different, inferior form of camera tube that he called an Iconoscope and a receiver, which he called a Kinescope, that was better than Farnsworth's.) This battle wasn't resolved until 1939, when federal courts ruled for Farnswoth, largely on the basis of testimony by his former teacher, Justin Tolman.

RCA grudgingly began paying royalties, but it was too late to make much difference. Two years later, production of TV sets was suspended for the duration of World War II; once the war ended—and just as TV was about to enter every home—Farnsworth's patents were set to expire.

In the meantime, electronic TV was much improved by a British research team led by Isaac Shoenberg, who, in 1936, demonstrated a new system that sharpened screen resolution to 405 lines, a vast increase in clarity over both the Farnsworth and Zworykin systems.

By the 1950s, television had become ubiquitous. Yet Farnsworth had become a figure of such obscurity that he could appear on the TV game show *I've Got a Secret* and stump the panelists who were trying to guess what he had invented. When one asked "Dr. X" if his device caused pain when used, Farnsworth replied, "Yes. Sometimes it's most painful."

Philo T. Farnsworth was denied much of the fortune and recognition due the inventor of the device that has shaped modern life. Yet in 1969, two years before he died, as he sat in his living room watching a man land on the moon on live TV, he turned to his wife and said, "You know, this makes it all worthwhile." ■

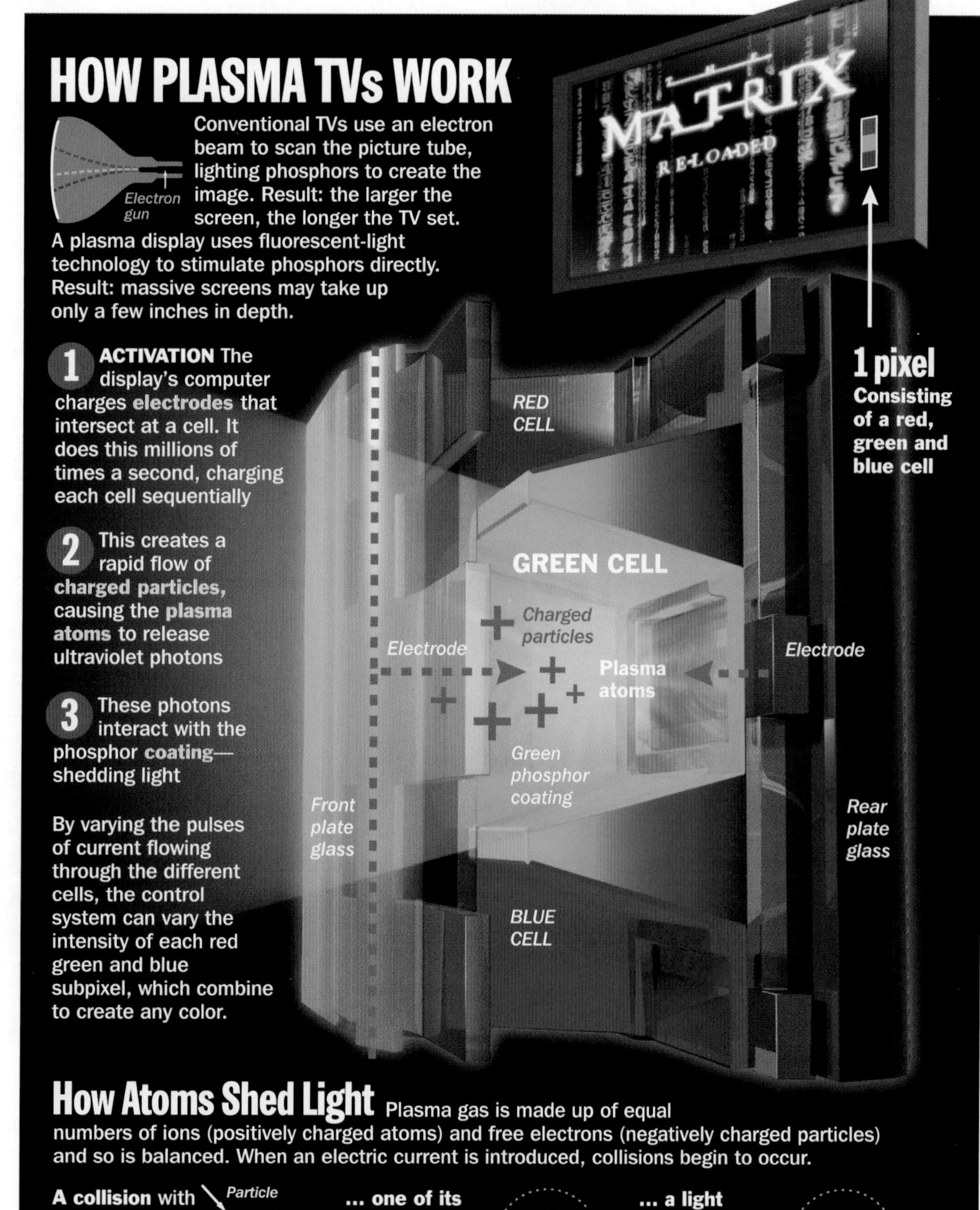

The Write Stuff

The typewriter transformed work, opened new doors for women and then became obsolete—all in just 100 years

TAPPED OUT
Sholes cashed out his patent for the typewriter too early, for only $12,000. At his death in 1890, the inventor was buried in a pauper's grave

CULVER PICTURES

On Dec. 9, 1874, Mark Twain wrote to his brother Orion Clemens, "I am trying to get the hang of this new-fangled writing-machine ... I saw the thing in Boston the other day and was greatly taken with it." The device in question—on which he tapped out those words—was a typewriter.

A gizmo to automate writing? The idea was age-old, but Twain bought one of the first practical machines to do the job. It was made by the Remington Sewing Machine Co. and based on a design by Milwaukee newspaper editor Christopher Latham Sholes and fellow tinkerer Carlos Glidden.

The pair also created the QWERTY keyboard, which aimed to deter jamming: not (as legend has it) by slowing typists down, but rather by placing letters that are often combined (such as *SH*) far apart on the keyboard.

A major drawback: the typist couldn't see what he (or she, for the

THE GRANGER COLLECTION

STARTER SET
Historian Richard N. Current called the first type-writer "a cross between a small piano and a kitchen table"

ROTARY SYSTEM
This early typewriter (the fifth to be patented in the U.S.) was slower than writing by hand and plagued by frequent breakdowns

HANSEL MIETH—TIMEPIX

Hands On

The modern pencil, filled with graphite (not lead), sheathed in cedar, painted yellow and topped with a rubber eraser, is a product of the mid–19th century. Massachusetts inventor Joseph Dixon improved upon a French process for mixing clay with graphite and tried to patent the idea of adding a small daub of tree resin from the West Indies to the tip so that mistakes could be rubbed out. (Rubber takes its name from this use.) A U.S. court ruled that Dixon's ideas were too basic to be patented.

French businessmen Marcel Bich and Edouard Bouffard acquired the rights to Hungarian inventor Ladislao Biro's ballpoint pen and launched the revolutionary Bic (a shorthand version of Bich's name) in 1950. The first ballpoint pen that was both reliable and affordable, it quickly became a major success.

BIC

BELOW: SCIENCE MUSEUM—SCIENCE & SOCIETY PICTURE LIBRARY

REINVENTING TYPING
Users of IBM's Selectric, right, could change fonts simply by replacing the rotating head. Below right, a virtual keyboard, introduced in 2002, can be projected onto a flat surface; its sensors read the strokes

typewriter offered a new, white-collar career to women) was writing, since the hammers and page were mounted on its back. The first "visible" type-writer—with paper facing the typist—was invented by John T. Underwood and Franz X. Wagner in 1898. Within three years, their model displaced the Remington as the most popular brand.

Improvements followed, yet the next great leap forward didn't come until 1961, when IBM released the Selectric, which eliminated hammers and the return carriage, replacing them with a golf-ball-size rotating head that moved back and forth across the page. By 1973, Big Blue's Selectric II offered a self-correcting feature. But by the mid-1980s, even IBM saw the typewriting on the wall: thanks to computers, typing was out, and "word processing" was in. The Selectric was discontinued, and today the typewriter has joined the buggy whip and the buttonhook on the antique-shop shelf, a charming curio of a bygone age. ■

1885: LINOTYPE
Form follows function in the linotype machine, which revolutionized printing by lowering the cost and time needed to compose pages of text

Gutenberg's Grandchildren

The challenge: putting words onto paper. The participants: inventors—of all types

CULVER PICTURES (2)

Four hundred fifty years is a long time to wait for the Next Big Idea. Yet in the late 1880s, printers had been using the same basic technology since Johannes Gutenberg invented movable type in 1436. Then three inventions revolutionized printing in only a few years.

The first was U.S. government clerk Tolbert Lanston's 1884 Monotype machine. Since Gutenberg, the metal letters that compose a printed page had been picked individually by hand and set in position on a plate. Tolbert's machine reduced this laborious task to just two steps. An operator with a keyboard typed the text, which perforated a paper tape with patterns of holes, each representing one character. Another operator fed the tape through a second machine, which "read" the tape and triggered brass letters to slide down from a bank into position on the plate. The machine automatically adjusted the spacing between letters to provide even margins ("justified," in printer's argot), and it produced finished plates four times as fast as those done by hand, with half the manpower.

A year later, Ottmar Mergenthaler, inset, invented the one-operator Linotype machine, which fused the individual letters into a solid line (or "slug") of molten lead. Dealing with an entire line of text as the basic unit of composition (rather than scores of loose pieces assembled into a line) was faster and produced more legible text. And the slugs could be melted down, recast and used again. The first Linotype machine went to work at the New York *Tribune* in 1886. Shortening deadlines, it made the news newsier; reducing labor, it made papers cheaper.

The same year, U.S. inventor Frederick E. Ives invented a way to reproduce photographs in print—the halftone process (below)—completing printing's second revolution, after Gutenberg. A third revolution came 80 years later: the photo-typesetting process, which creates type by exposing film onto photosensitive paper, was developed in the early 1960s.

The fourth revolution, digital printing, ramped up in the 1980s. It has made each of us a Gutenberg and every desktop a printing plant—as long as that evil laser printer doesn't jam. ■

HULTON ARCHIVE—GETTY

SLUG SHOP
Linotype operators set text at *Lloyd's Gazette* in Britain, 1939. Inventor Ottmar Mergenthaler was born in Germany and immigrated to the U.S.

HOW WE PRINT PHOTOS

In 1886 U.S. inventor Frederic E. Ives solved a problem that had bedeviled printers since the invention of photography some 50 years before: how to print black-and-white pictures on a page, capturing the full range of their various shades of black, white and grey. Ives' process involves converting the image into a series of miniature black and white dots, which, like tiles in a mosaic, simulate a continuous image to the eye, with full ranges of grey. In Ives' process, these "halftones" were produced by taking a picture of the original photograph through a screen. Today photos are scanned into computers, and the halftone screens are created digitally.

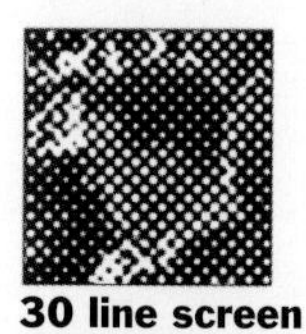
30 line screen

60 line screen

90 line screen

120 line screen

PRINTING IN COLOR

To print full color images, four color inks are used: cyan (blue), magenta (red), yellow and black. Each of these colors is screened and printed—lightest to darkest—on the printing press. As with the black-and-white halftone process, the colors combine to create a true-to-life image. To avoid image-skewering moiré patterns, each screen is tilted at a slightly different angle.

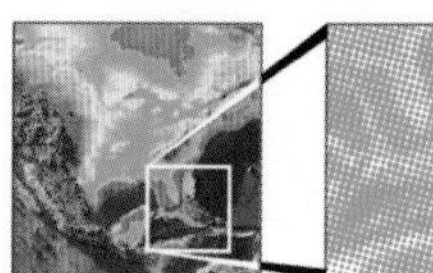

Color image

Cyan

Magenta

Yellow

Black

Traditional screen angles used to minimize moirè patterns

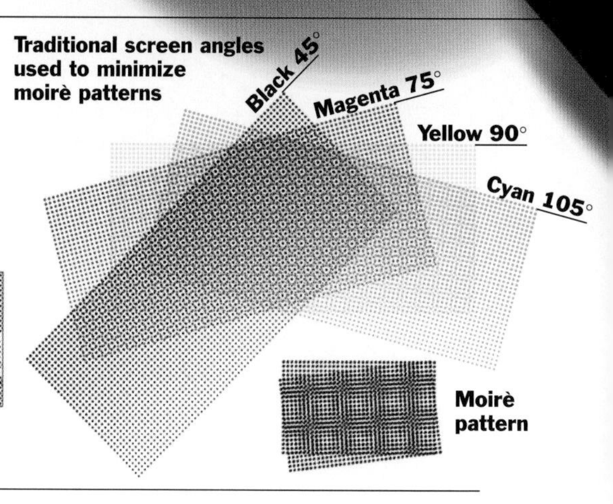

To Write with Light

A spectrum of glowing gases created a new kind of sign language

Late in the 17th century, scientists discovered that air could be made to glow when sealed in an airtight glass tube under low pressure and shaken. But as no one understood the science behind the phenomenon, it was regarded as a laboratory novelty. Then in 1855, German glassblower Heinrich Geissler solved the mystery when he ran electric current through one such tube and achieved the same effect. The glow was electricity made visible—in the form of a static charge generated by the shaking.

Enter Georges Claude, a French inventor seeking a cheap, quick process to extract oxygen from the air for use in hospitals and welding torches. His experiments yielded by-products, large amounts of the six "noble" gases (so-called because they do not easily form chemical bonds with any other element): helium, argon, krypton, xenon, radon and, predominantly, neon. When Claude filled glass tubes with these gases and zapped them with electric voltage, a marvelous spectrum emerged. Each tube glowed a different color: neon shone orange-red, argon burned lavender, helium gave off a flesh-colored hue, and so on.

At first, Claude hoped he had found a competitor for Edison's electric light bulb. But neon light (as Claude's creation came to be called, even though he used a variety of gases) could not match Edison's incandescents in cost, luminosity and fidelity to natural light.

So Claude remade his discovery: neon would be used not to illuminate but to communicate. The first neon sign was installed by Claude's company at Paris' Grand

©TOPHAM—THE IMAGE WORKS

BROWN BROTHERS

LIGHT SHOW
Georges Claude displays a gas light at the Academie Francaise. Above, Las Vegas got its first major neon sign in 1954; today the Strip's lights can be seen by astronauts in orbit

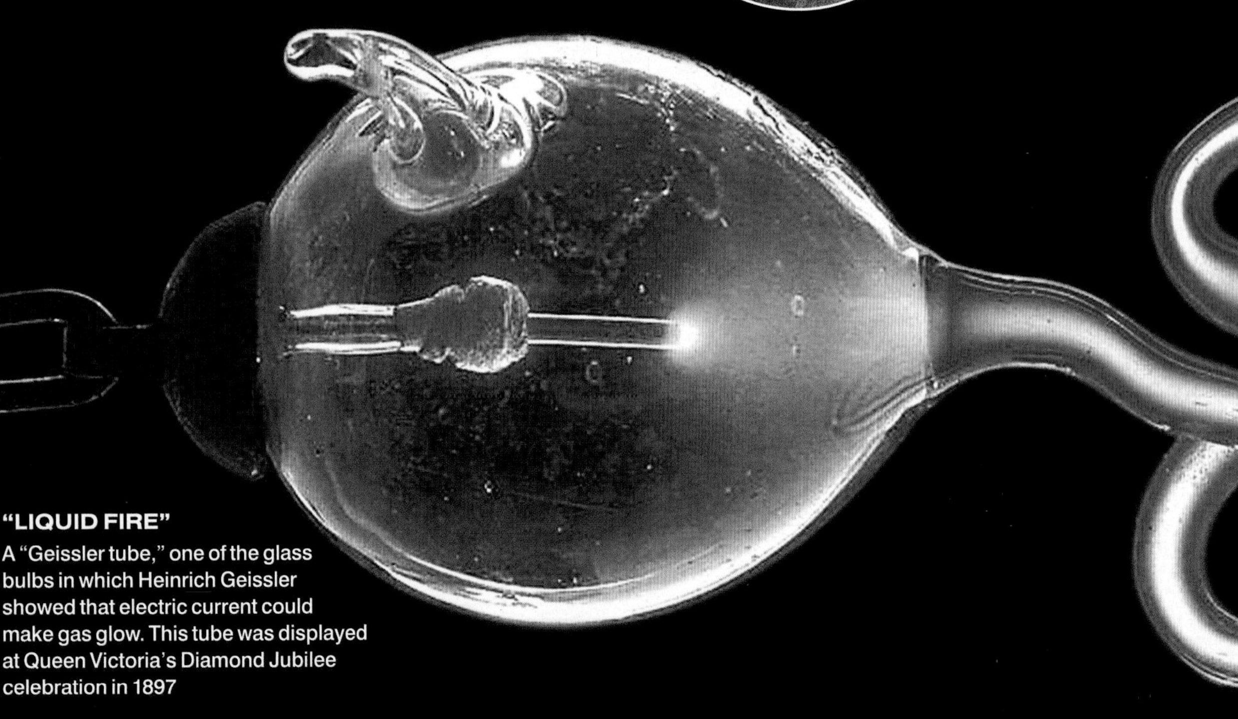

©SCIENCE MUSEUM/TOPHAM—THE IMAGE WORKS

"LIQUID FIRE"
A "Geissler tube," one of the glass bulbs in which Heinrich Geissler showed that electric current could make gas glow. This tube was displayed at Queen Victoria's Diamond Jubilee celebration in 1897

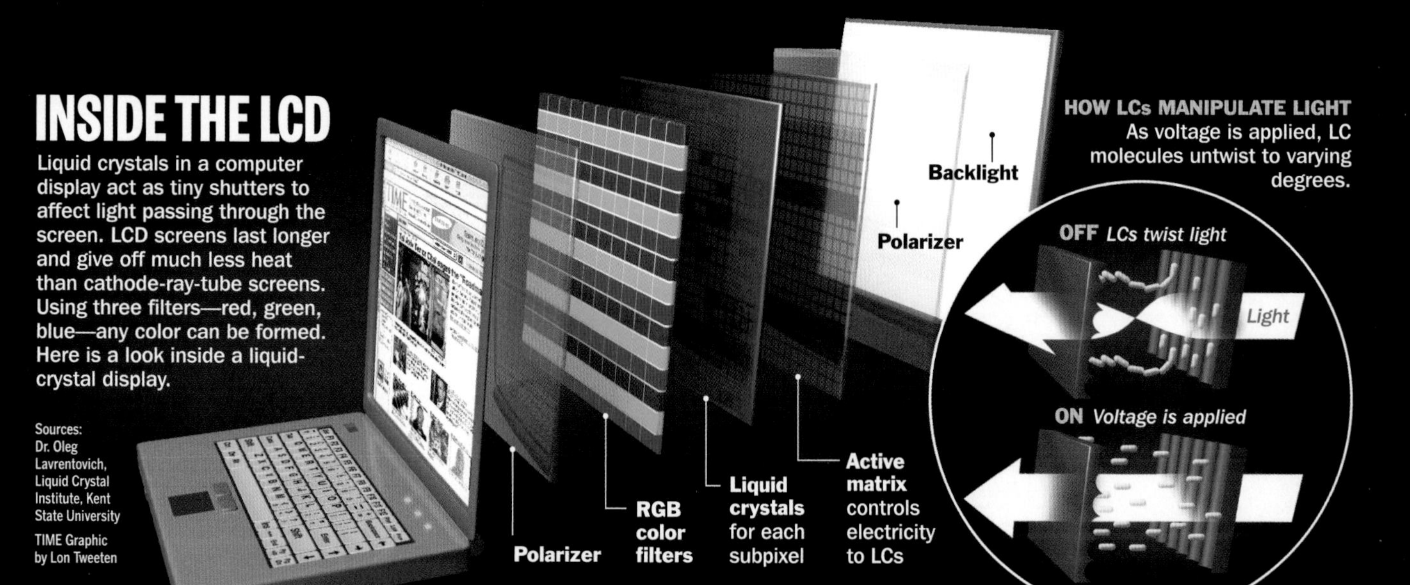

Palais in 1910. Next came a barbershop in Montmartre, then a huge sign for the Paris Opera House. Neon came to America in 1924, when a Los Angeles Packard dealership commissioned two red-and-blue signs (cost: $24,000). The "liquid fire" actually stopped traffic on Wilshire Boulevard.

Along the way, Claude invented an electrode that could work for hundreds of hours without heating up, sputtering or burning out, making neon lights last longer with less maintenance and paving the way for fluorescent lighting. Today high-tech light technologies—liquid-crystal displays and light-emitting diodes—are essential tools of modern life. ■

Pathways of Light: Fiber Optics

By the late 1960s, telephone engineers were eager to replace the pair of copper wires that had been carrying phone calls since the days of Bell. Copper's proposed stand-in: glass fiber. Impervious to the water and lightning that bedevil copper wires, and a more efficient conductor than metal since it is not subject to electrical resistance, a fiber-optic cable thinner than a strand of human hair can handle hundreds of thousands of phone conversations, data transfers and video feeds simultaneously—saving expense, space and maintenance.

In 1970, Robert D. Maurer, Donald B. Keck and Peter C. Schultz, a team of researchers at Corning, found a way to combine fused silica and titanium in a fiber-optic line that resulted in minimal leakage of signals. Their invention made possible superior long-distance phone calls, high-speed computer networking (such as the Internet) and new medical technologies. Some 90% of all U.S. long-distance phone calls now travel over fiber-optic lines.

EYEWIRE—GETTY

SNARED IN THE NET
The World Wide Web celebrates its 10th anniversary in 2003. One of history's most rapidly evolving technologies, it has reshaped the way we communicate, learn, do business, get the news, trade stocks—you name it. Below, cyberspace voyagers clog the lines at an Internet café in New York City's Times Square

NINA BERMAN—AURORA

Nothin' but Net

How the cold war gave birth to today's hottest technology

Al Gore was accused of having claimed he invented the Internet, but Nikita Khrushchev may deserve more of the credit. The launch of Sputnik in 1957 spawned many U.S. catch-up programs—among them the formation of the Advanced Research Projects Agency (ARPA), a Department of Defense office charged with ensuring that the U.S. maintained a strategic lead in any technology involving national security.

One of these advanced projects was an experiment to link mainframe computers at disparate locations so that users (mostly researchers at universities doing defense work) could share information electronically. Doing that required turning computer data into telephone sounds. The invention that did it, the modulator/demodulator (modem) was developed by AT&T scientists in the late 1950s. (A commercial version went on sale in 1962.) The first experimental computer link spanned the nation, connecting M.I.T.'s Lincoln Lab and Santa Monica's System Development Corp. in 1965. Once the idea proved feasible, the Pentagon began rolling out "ARPANET" in 1969.

Two years later, a researcher in Cambridge, Mass., Ray Tomlinson, wrote the first e-mail program, choosing the @ symbol to separate the name of a user from the network on which he or she worked. By the late 1970s, several thousand users (still mostly academics) were "online," establishing communities, or newsgroups, to share interests (the first one, SF-Lovers, was about science fiction; San Francisco's

Webmasters

Tim Berners-Lee, above, poses with his creation, the first page to appear on the World Wide Web. The idealistic Berners-Lee, an Oxford graduate, has steadfastly refused to cash in on his invention, which he first envisioned as a program called Enquire in 1980. In contrast to Berners-Lee is American Marc Andreessen, left, who helped create the first graphic Web browser, Mosaic, as an undergraduate. He later made hundreds of millions of dollars when his improved browser, Netscape, became a stock-market phenomenon in 1995.

CERN

WILLIAM MERCER MCLEOD

influential Whole Earth 'Lectronic Link, or WELL, followed). In the meantime, ARPANET was cut loose from the Pentagon. Now known as the Internet, it was supervised by the National Science Foundation.

Six years later, a British researcher at the CERN nuclear facility in Switzerland, Tim Berners-Lee, made the Internet user-friendly for all when he created a program, Hyper Text Markup Language (HTML), that made it possible to post text, pictures and sound on what he called the "World Wide Web."

Internet activity took its next giant step in 1993, when Marc Andreessen and a group of student programmers at the University of Illinois cobbled together the first program that made Web pages easily viewable in graphic form. They called their "browser" Mosaic. Two years later, the same team launched an improved browser, Netscape, whose initial public stock offering touched off a speculative mania that gripped U.S. financial markets for the rest of the decade.

Cyberstocks fizzled in 2000, but the Internet boom continues, driven by the migration of its messages from phone lines to far-faster cable—and next, it seems, into wireless form. Once a plain-vanilla world of black text on gray pages, the Internet is now pulsing with brilliant colors, buzzing with news updates, swarming with "instant messages" sent by teens. Ten years after the Web was first spun, the Internet has joined the telephone and the TV as an essential tool of communication. And they say the Russians never invented anything. ■

How We Record

Edison Cylindrical Phonograph, 1877

4

The Unblinking Eye

Artists' palette and scientists' tool, family historian and sweetheart's friend, the camera is a marvel of technology that still seems magical at its core

French writer Charles François Tiphaigne de la Roche described a photograph as "a most subtle matter, very viscous, and proper to harden and dry, by the help of which a picture is made in the twinkle of an eye ... This impression of the images is made the first instant they are received [and] immediately carried away into some dark place. An hour after, the subtle matter dries and you have a picture."

This is not a somewhat flowery attempt to explain modern photography: it is the fancy of a European science-fiction novelist writing in 1760. One reason De la Roche was able to picture photography so precisely in his *Giphantie* (the title is an anagram for one of his names) is that its twin technologies—the optical process of capturing an image in a box and the chemical process of recording patterns of light in photosensitive dust—had been known for many years. The problem was simply that no one had connected them. Photography is one of those rare inventions that didn't demand original technology so much as an original vision, a new way of assembling existing ideas.

The man who did so was French army officer Joseph Nicephore Niepce. On a summer day in 1826, he combined a camera obscura (a device invented in the 1500s) with a plate

MAN OF METAL
A daguerreotype of ... M. Daguerre. His images, exposed onto metal, were the sharpest and clearest of their day but could not be copied or printed

MAN OF SILVER
Henry Fox Talbot's silver calotypes are considered the first art photos. Later in life he abandoned his camera to decipher ancient Assyrian texts

PORTRAIT OF THE ARTIST
Photo pioneer Talbot, standing at right, demonstrates his camera and its revolutionary silver-paper film process at his studio in Reading, England, in 1868. The first photographer to use a negative, Talbot was able to make endless copies of any picture he shot

Eastman's Brownie

Named for its designer, Kodak camera genius Frank Brownell, and priced at $1 (plus a dime for a six-exposure roll of film), George Eastman's Brownie was bought by more than 150,000 customers in 1900, its first year on the market. By 1920, more than 8 million Brownies had been sold. Eastman (photographed above holding an original Kodak camera aboard ship by artist Frederick Church in 1890) drove his company to create the simple device, his vision of cameras for all anticipating Henry Ford's dream of autos for all.

A cultured man of wide interests, Eastman was a patron of the arts and a philanthropist. Yet this famous, vastly wealthy man—whose work allowed so many faces to be captured forever—was so retiring that he could walk the streets unrecognized. Afflicted with growing ailments, he took his own life at 77 in 1932.

HENRY GROSKINSKY—TIMEPIX

covered in light-reactive material (which had been discovered about 100 years earlier). Before that year, the camera obscura, a box with a small opening at one end in which a convex lens is mounted, had been used mainly by landscape painters. Incoming light would project onto a glass screen—in perfect scale and perspective—the image of whatever was in front of the lens. But instead of tracing the image inside the box, as painters did, Niepce slid a pewter plate coated with bitumen behind the glass. Then he left the camera obscura (a Latin term meaning "darkened room") near the window of his country home in Provence. As it baked in the sun for eight hours, the coating absorbed a pattern of light and darkness. The light areas of the bitumen hardened; the dark areas remained soft. At dusk, Niepce washed off the unexposed, still-soft portion of the coating with lavender oil. What remained was a hazy silver-and-black image of the pear tree in his courtyard: the first photograph.

Niepce soon partnered with painter and theatrical designer Louis-Jacques-Mande Daguerre. After Niepce's death in 1833, Daguerre replaced the

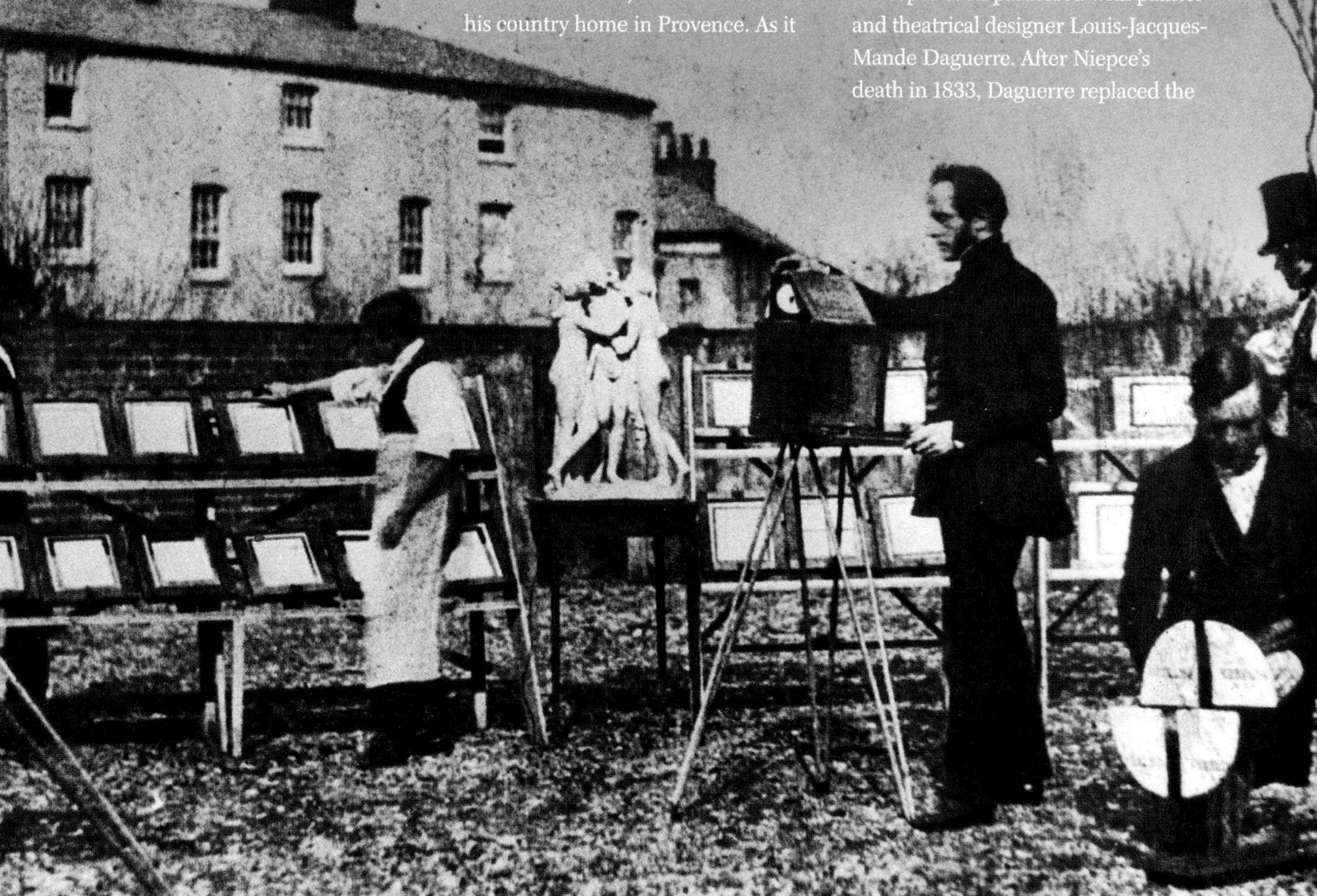

1904: AUTOCHROME

Pioneers of cinema as well as photography, the Lumière brothers developed the Autochrome system, based on potato flour and dyes. In this 1913 still image, the model is wearing red because it is the color best captured in Autochrome

COLOR FILM

The Lumière brothers' Autochrome film, below, was improved upon in 1935. Kodak chemists Leopold Mannes and Leopold Godowsky applied the principle of subtractive color (in which blue, e.g., is created by filtering yellow out of green) to 35-mm camera film and came up with Kodachrome, the first modern film stock for color stills

bitumen with mercury, which absorbed light much faster and could expose a picture onto metal in 20 minutes, rather than eight hours. Daguerre patented this process in 1839, and the revolutionary new medium soon became wildly popular.

But each "daguerreotype" was literally one of a kind: it could not be printed or duplicated. Two years later, Englishman William Henry Fox Talbot substituted a mixture of salt and silver nitrate for Daguerre's mercury and perfected a new system for exposing photos onto paper, rather than metal. Talbot, who called his new kind of image a calotype (Greek for "beautiful picture"), also invented the seminal process of shooting negative images, in which light and dark areas are reversed, and from which an unlimited number of positive duplicates can be made.

Yet even as photography progressed, it remained expensive and cumbersome, the province of professionals. Its full promise wouldn't be realized until a few essential adjectives could be applied to it: affordable, colorful, convenient, fast.

A young banker from upstate New York tackled the first challenge. George Eastman believed every family should have a camera, so cumbersome glass plate negatives (which had to be hand-coated in emulsion before being exposed) had to go. With no background in chemistry, Eastman experimented until he developed a dry cellulose film, superior to glass.

Eastman's first camera was unveiled in 1888. The Kodak (an invented word) cost $25—at the time, a rather hefty sum—and contained film for 100 exposures. When these had been shot, the camera was mailed to Kodak, which developed the film (for $10), loaded a new roll in the camera and mailed it back to the customer. Kodak sold 900,000 of these cameras in 12 years, until 1900's revolutionary Brownie, a $1 camera, fully realized Eastman's dream of affordable photos.

The next step: color. In 1904, French brothers Auguste and Louis Lumière pioneered this area with their Autochrome system. Although the photos appeared on glass, could not be duplicated and required a special viewer, a milestone had been reached: photographs captured life in color.

Now cheap and capable of working in color, cameras were still bulky. In 1924, German craftsman Oskar Barnack adapted the motion-picture standard of 35-mm film to still photography, introducing the world's first lightweight precision camera, which would become famous under the name Leica. Barnack, an outdoorsman, was inspired to miniaturize cameras because his asthma kept him from toting his bulky model on hikes.

The final step was to eliminate the darkroom. Inventor Edwin Land came up with the configuration of negative film, positive paper and developing chemicals to produce instant photos in 1948. Fifteen years later, he did it again—this time in color.

Its four adjectives achieved, photography was poised for its most recent incarnation, as a digital medium. Yet its essence remains as De la Roche envisioned it in 1760: "A picture so much the more the valuable, as it cannot be imitated by art nor damaged by time." ■

1924: LEICA

Barnack named his downsized camera by joining the first three letters of his employer's name (Leitz) with the first two letters of the word camera

LEICA (X2)

LAND'S INSTANT CAMERA

This 1972 Polaroid features an ingenious folding body, allowing it to fit into a coat pocket. A motor automatically ejected finished color images out of the camera

GEORGE EASTMAN HOUSE

"Doc" Edgerton's World

The first modern flash bulb was invented in 1924 by Austrian engineer Paul Vierkotter. His magnesium-coated wire in an evacuated glass globe replaced the dangerous powder flashes that had been standard for decades. Advancing this line of development, electrical engineer and M.I.T. professor Harold (Doc) Edgerton invented the strobe lamp in 1931, then turned his camera on running water, speeding bullets, hummingbirds in flight and many other subjects. Exposure times of one-millionth of a second and less allowed his camera to unlock events that ordinarily happen too fast for human perception to register. Although Edgerton's works are science, their striking beauty led many to collect his images as art.

ALFRED EISENSTAEDT—TIMEPIX

POW!

Hammer meets bulb, as Edgerton's strobe-equipped camera saw it

HAROLD EDGERTON—PALM PRES

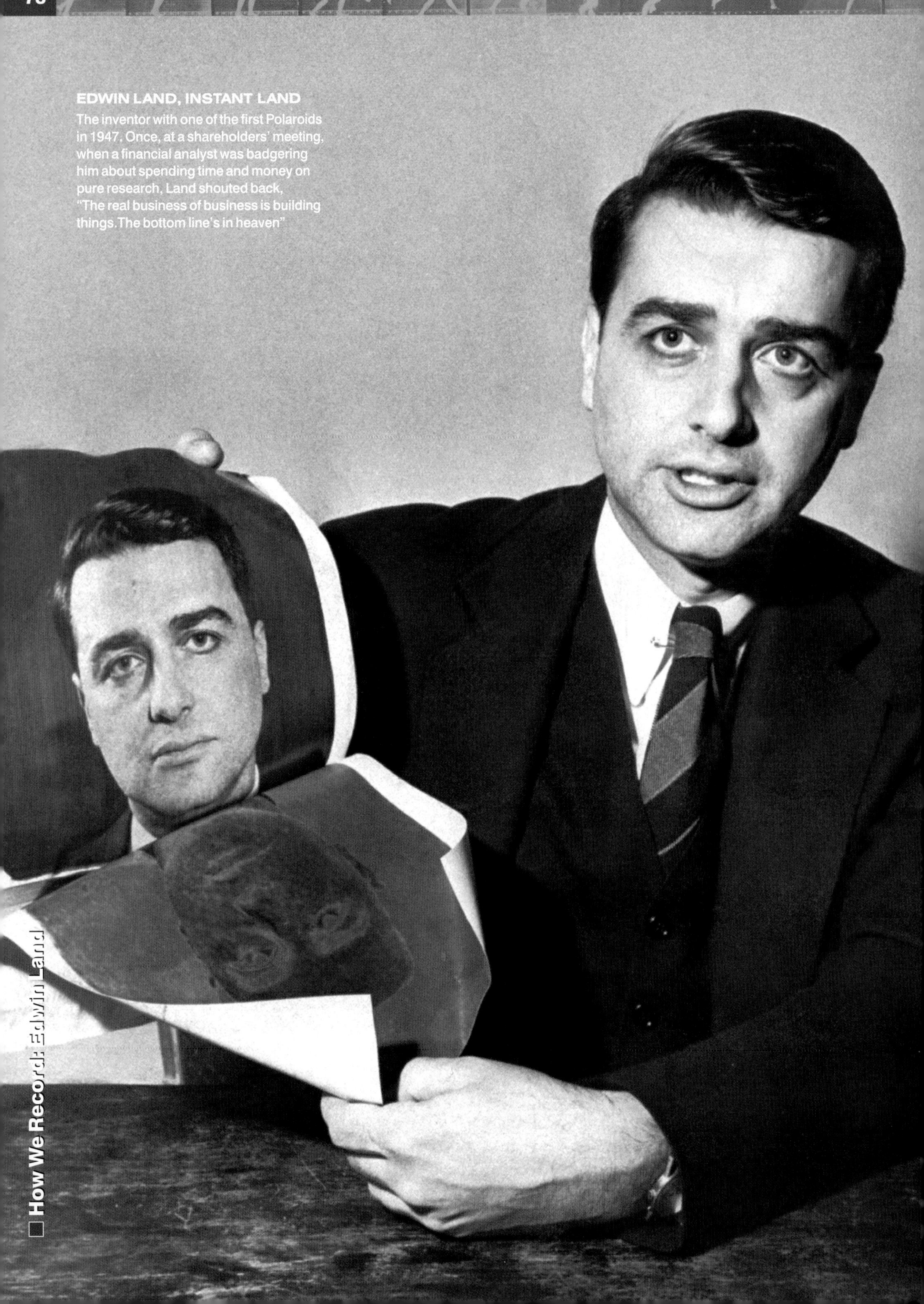

EDWIN LAND, INSTANT LAND

The inventor with one of the first Polaroids in 1947. Once, at a shareholders' meeting, when a financial analyst was badgering him about spending time and money on pure research, Land shouted back, "The real business of business is building things. The bottom line's in heaven"

Profile

Double Vision

Invention often requires looking at the world in new ways. Edwin Land reversed that polarity: he invented new ways to look at the world

HARVARD FRESHMAN EDWIN HUBBLE LAND WAS walking the streets of Cambridge at twilight when he was irritated, not for the first time, by the glare of headlights from onrushing cars. Land, who had been fascinated by optics and light since his high school days, began pondering why a car's lights had to shine so brightly that they made onlookers shield their eyes and squint. Within days, he formulated an idea based on a fact he had read years earlier, as a precocious Connecticut schoolboy: light waves, which usually shimmer randomly in all directions, can be arranged in more orderly fashion by certain transparent crystals. The effect is called polarization, and Land realized that if these large, expensive crystals could be downsized and mass-produced on cheap sheets of plastic, they might have enormous practical value.

But he lacked a lab. So Land began breaking into a Harvard facility in the middle of the night to conduct experiments. Once he knew he was onto something, he dropped out of college and moved to New York City. Within 18 months, he filed a patent for his new light-polarizing material. In 1932, he founded his company, which became the Polaroid Corp. in 1937.

Inexpensive, thin and transparent, Polaroid's plastic filter could be cut to any size and soon found uses in sunglasses, ski goggles and cameras—though the automotive industry shunned it. By the mid-1930s, Land had become famous as a brilliant inventor. During World War II, the U.S. military called upon him to invent dozens of new technologies: an optical sighting system for tank cannons, plastic lenses for gun scopes, a range and elevation finder to target enemy planes.

During the war, on Christmas vacation with his family, Land took pictures of his 3-year-old daughter Jennifer. "Why can't I see them now?" she asked. Land went for a long walk to cogitate; he later wrote that "within an hour, the camera, the film and the physical chemistry became clear to me."

Land envisioned a radically new process: exposing the film and then developing the negative and the print together, using a system of rollers to squeeze the exposed film up against the print paper. As the rollers moved along the negative and the paper, they would also break open a small sealed vial, which contained the developing chemicals. The rollers would then spread the chemicals between the negative and the paper.

> "Within an hour, the camera, the film and the physical chemistry became clear to me."

It took Land a single afternoon to dream up the process—and four years of hard work to perfect it. In 1948, his Model 95 Polaroid-Land camera went on sale in Boston. By 1956, Land had sold more than 1 million of them. In 1972, he unveiled the SX-70, a technically sophisticated camera with complex optics that allowed it to take instant pictures that rivaled the clarity and rich colors of images developed in a darkroom.

Land was a driven, intensely private man; he was querulous, imperious, unpredictable and impatient. After a cascade of successes, he stumbled in the late 1970s when he unveiled a long-cherished dream. His instant movie camera, Polavision, was technically brilliant but late to market. Outpaced by video recorders, Polavision foundered, Polaroid stock tanked, and in 1982 Land was forced out of his own company.

The inventor went on to found the Rowland Institute, a Cambridge laboratory dedicated to the pure research that had always been his first love. The Polaroid Corp. spiraled into a long decline. Still, when he died in 1991, Land held 535 patents, second only to Thomas Edison—pretty good company for a former lab burglar. ■

Eadweard Muybridge

CULVER PICTURES

Reality on Sprockets

Photographs on overdrive, the motion picture started out as a tool of science and ended up as a storyteller

Even for a 19th century railroad tycoon like Leland Stanford, $25,000 was serious money. Yet that was the amount the former California Governor had riding on a bet with a fellow swell from San Francisco's Nob Hill. The wager: Stanford maintained that a galloping horse lifts all four of its legs off the ground simultaneously with each stride, becoming airborne so briefly that human perception can't register it. So in 1872, Stanford hired photographer Eadweard Muybridge—the English-born Edward Muggeridge, who had changed his name to something a bit more exalted—to prove him right.

Muybridge experimented for a year with shutter speeds, lighting techniques and film stocks until he was able to reduce the typical exposure time for still photographs from several minutes to several hundredths of a second. In 1873 he set up a series of cameras at ground level on the track at Stanford's Palo Alto farm and strung trip wires across the horse's path. When Stanford's favorite trotter, Occident, galloped down the track, it triggered the shutters of each camera in sequence. The photo strips that

GRANGER COLLECTION

MANSELL COLLECTION—TIMEPIX

1895: CINÉMATOGRAPH

The first public demonstration of the Lumière brothers' device. Both camera and projector, the unit's film strips lasted much longer than did the Kinetoscope's

1885: MUYBRIDGE MOTION STUDY
Muybridge's camera took this sequence of a gymnast's somersault, while another camera shot the same event head on

1893: MOVIE STUDIO
Thomas Edison's "Black Maria" was his film research lab. Built on a revolving base (note wheels at right), it followed the sun's path to capture maximum natural light

resulted won Stanford his bet and did something more: they captured motion on film for the first time.

For nearly a century, nickel-paying customers had been peering through viewers and turning the hand-cranks on such devices as the Praxinoscope and the Zoetrope, which created the illusion of motion by rapidly flipping cards containing sequential drawings. But Thomas Edison was among the first to think of using this technology to display a series of photographs (shot in sequence, like Muybridge's) and mounted on a continuous strip of celluloid, rather than on separate cards.

George Eastman's new Kodak company supplied the film stock, and Edison's assistant, William Dickson, designed and built the first Kinetograph, a specialized camera that took rapid-series pictures. They were viewed in a Kinetoscope, a hand-cranked box that would show a back-lit 50-ft. loop of film, lasting less than 30 seconds. But these visions were still viewed by one person at a time through a peephole.

Enter Auguste and Louis Lumière, brothers who had inherited their father's photography studio in Lyons,

NMPFT—TOPHAM-HIP—THE IMAGE WORKS

1894: EDISON KINETOSCOPE
Viewers saw only 20 seconds of film, through a tiny aperture

Animation

Walt Disney didn't invent animation in films—even the jerky illustrations seen in hand-cranked Zoetrope machines are clear progenitors of Mickey Mouse & Co. But Disney and his colleagues certainly were pioneers of many of the format's defining technologies. In 1928, Disney's *Steamboat Willie* merged the day's new rage—talking pictures—with animation. Disney's studio later invented the four-layer camera, seen at right, which achieved new richness of depth in filmed cartoons. *Toy Story* (1995) was the first animated feature created not with a hand-drawn, stop-motion process but on computers.

J.R. EYERMAN—TIMEPIX

WARNER BROTHERS

1927: "THE TALKIES"
Warner Brothers' *The Jazz Singer,* with Al Jolson, wasn't the first sound film; it was second to Warner's *Don Juan*—and was largely silent, with brief sound segments

France. The Lumières realized that by using a powerful lamp, they could project images from celluloid film onto a large screen, with far more impact than the Kinetoscope's peephole provided. In 1895 they exhibited the first motion pictures shot with their Cinématograph, a hybrid camera-projector, in Paris. When a locomotive seemed to come speeding out of the wall, some of the viewers ran into the street screaming.

Within a few years, the motion-picture industry had taken shape—but

SUPERCOLOSSAL CINEMA

Always fiddling with new film formats, desperate movie moguls reached into their bag of tricks when TV arrived in the '50s. Three of these formats proved faddish. IMAX, from the '70s, has proved its staying power.

Cinerama

Invented by film engineer Fred Waller, Cinerama achieved a widescreen effect by synchronizing three standard 35-mm filmstrips to create a single huge image—if one a bit rough at the seams—on a curved screen. *This Is Cinerama,* a movie made to showcase the format, proved a sensation when it opened in 1952. Seven-channel stereophonic sound—also an innovation—helped the buzz. Today there are only three theaters in the world configured to show Cinerama films.

CinemaScope

Impressed with Cinerama's success, Hollywood studios raced to Paris to buy rights to a system patented by Prof. Henri Chrétien in the late 1920s, which achieved its widescreen effect with an anamorphic lens. The lens squeezed the image during filming, fitting a wide picture on the 35-mm filmstrip; another lens in the projector unsqueezed the image on the screen. The first film shot in CinemaScope, *The Robe,* was released in 1953.

mainly as a producer of novelties like Edison's short film, *The Sneeze,* or the travelogues that became a specialty for the Lumière brothers. It took Edwin S. Porter, another Edison cohort, to realize film's dramatic potential, using storytelling and the trick—uniquely suited to cinema—of cutting from one scene to another to excite audiences. His 1903 film *The Great Train Robbery* dramatized in 14 scenes and 10 minutes the story of an actual heist committed by the infamous Wild Bunch in Wyoming three years before, thus becoming both the first true dramatic film and the first western.

Film was now a storyteller—a mute one. In the 1920s audiences gathered to watch movies without sound, then went home to listen to radio shows without images. The need to unite the two was clear. In 1920 radio pioneer Lee de Forest invented PhonoFilm, an optical system in which sound waves were imprinted as shades of light and dark on movie film, which perfectly synchronized audio with the pictures onscreen. Hollywood thought it knew better, though, and initially adopted Vitaphone, a clumsy system using a phonograph connected to the projector. Early sound films, like 1926's *Don Juan* and 1927's *The Jazz Singer* were big hits, and in the mid-'30s, the industry dropped Vitaphone in favor of an optical system based on De Forest's work—just as his patents expired.

The movies' final hurdle was to capture color on film stock. Inventor Herbert T. Kalmus spent 20 years perfecting his Technicolor process—which uses three film strips—red, blue and green—to create full color. He spent another decade convincing studio executives to try it. Finally, Walt Disney agreed to use Kalmus' invention for his 1932 animated short, *Flower and Trees.* But it wasn't until 1939, with the release of *Gone With the Wind* and *The Wizard of Oz,* that color's possibilities began to be fully exploited.

After World War II, movies were threatened by a new medium, television, and the industry tried to reinvent itself with wider screens and even 3-D glasses. Not to worry: the love of congregating in a dark theater and watching stories told by projected shadows and light seems to be universal. Auguste Lumière said in 1895, "Our invention can be exploited for a certain time as a scientific curiosity, but apart from that it has no commercial future whatsoever." *Au contraire:* in 2002, human beings spent $20 billion dollars to watch the movies—a pretty good return on a $25,000 bet. ■

PHOTOFEST

1935: TECHNICOLOR *Becky Sharp,* starring Miriam Hopkins, was the first full-length feature film shot on the new system

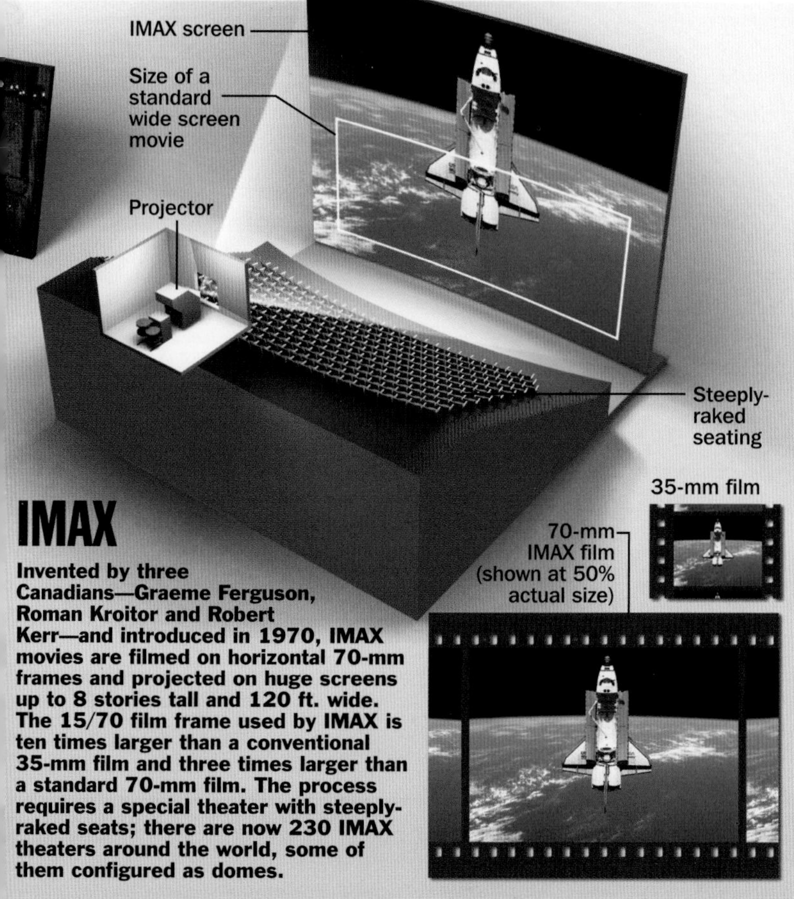

IMAX

Invented by three Canadians—Graeme Ferguson, Roman Kroitor and Robert Kerr—and introduced in 1970, IMAX movies are filmed on horizontal 70-mm frames and projected on huge screens up to 8 stories tall and 120 ft. wide. The 15/70 film frame used by IMAX is ten times larger than a conventional 35-mm film and three times larger than a standard 70-mm film. The process requires a special theater with steeply-raked seats; there are now 230 IMAX theaters around the world, some of them configured as domes.

3-D Movies

3-D effects in both print and movies are created by sending slightly different images to each of our eyes, providing the paralax view that supplies a sense of depth to a flat surface. Hollywood's first 3-D movie was 1953's *Bwana Devil.*

Black-and-white 3D images can be viewed with anaglyphic glasses. These have one red and one green lens, each lens masking out one image.

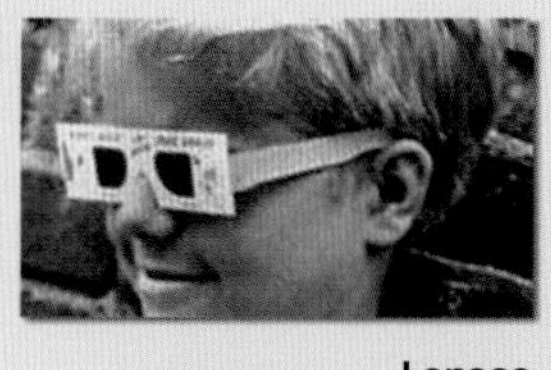
BETTMANN CORBIS

Color images can be projected and viewed with polarized light. One lens blocks light that is polarized horizontally, the other blocks light that is polarized vertically. The film is shot with two side-by-side polarized lenses.

Liquid-crystal glasses use electrically activated lenses to block each eye alternately, wirelessly synchronized with a stereoscopic image on the screen that is beamed by two projectors.

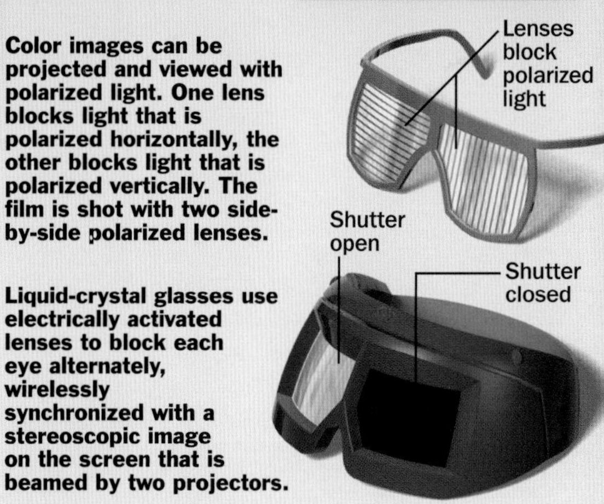

TIME Graphic by Joe Lertola Sources: IMAX® Corp.; www.widescreenmuseum.com

BROTHERS, PARTNERS

"The cinema is an invention without a future," Louis said more than once. The two inventors, above in 1895, walked away from their creation early in its life

Profile

First Family of Film

Movies were tiny, brief, private visions viewed through a peephole. The brothers Lumière created a shared, larger-than-life sensation

OUTSIDE THE GRAND CAFÉ ON PARIS' BOULEvard des Capucines, a nervous Antoine Lumière hung a hand-painted sign bearing the words "Cinématographe Lumière—One Franc," with an arrow pointing toward the café's basement. There, in a former billiards parlor, he ushered the 33 paying guests (a disappointing turnout; he had rented more than 100 chairs) to their seats. It was the evening of Dec. 28, 1895, and Antoine, the father of Auguste and Louis Lumière, was preparing for the first public exhibit of his sons' new invention, which they had named by fusing the Greek words meaning "to write with movement."

That evening, Lumière showed 10 films that his sons had made over the previous few months, each slightly less than a minute long. The audience laughed at the sight of a man repeatedly trying to mount a horse; they were touched to see Auguste's baby daughter; they were terrified by a train rushing toward them. They had witnessed the birth of what the French call the seventh art: moving pictures projected onto a large screen to be watched by many people at once. It was an enormous advance beyond Thomas Edison's version of Kinetoscope movies—small, short, private, peephole affairs—and it quickly transformed motion pictures from a novelty into a major industry.

The secret behind the sorcery? Louis came up with the idea of using a mechanical claw (powered by a hand crank) to advance the roll of film inside the camera by one frame, hold it in place while the shutter opened, and then move the next frame into position while the shutter was closed. The lightweight camera they designed for this task was then modified to serve as a miniature darkroom (the film developed inside over the course of several hours), then redesigned again (a powerful lamp was added) to function as a projector.

From that first picture show, word spread quickly across Paris (and then around the world) that technology had given birth to an entirely new form of art. Within days, the lines outside the Cinématographe Lumière stretched for a quarter of a mile, and receipts swelled to more than 2,000 francs per day. They were soon fending of large offers of cash from Edison and many others who saw unlimited potential in what the two had created. The Lumières quickly parlayed their initial success into a substantial business by sending camera crews around the globe to produce *actualités* (their term for the short travel documentaries that became their specialty) about every major city in the world. And then, without warning,, they turned their backs on movies. Louis Lumière would later explain that "I quit because I saw the artist coming into the industry. Georges Méliès was a genius," he said, referring to the innovations in special effects and narrative form introduced by Méliès, a former magician, in movies such as *A Trip to the Moon.* "I was not an artist. I was not a filmmaker."

> **"I quit because I saw the artist coming into the industry. I was not an artist. I was not a filmmaker." —Louis Lumière**

He was, however, an inventor. Continuing to work together, Louis and Auguste later developed Autochrome, the first practical color film for still cameras. On his own, Louis patented a new design for audio speakers, while Auguste later created a sterile, nonstick bandage that is still sold today. And to the end of their lives, Auguste and Louis Lumière kept a promise they had made to each other as small boys: that whatever they invented, they would always share as equal partners. It's nice to reflect, as the houselights dim and the crowd hushes, that the movies, a beloved communal art form, were born of brotherly love. ■

GENIUS AT WORK
Edison, right, and his younger colleagues at work in his New Jersey laboratory. Edison always insisted that his first recording was of the nursery rhyme, *Mary Had a Little Lamb*, but at least one biographer says that Edison's actual first words into the phonograph were a joke about a stingy Scotsman

The Sound of Music

Don't tell Eminem, but it all began with Mary Had a Little Lamb

As he turned 30 years old, Thomas Edison was not yet Thomas Edison. He was a successful scientist and businessman, to be sure, but he had not yet become the presiding genius of history's greatest age of invention. That Thomas Edison came to life on Nov. 20, 1877.

One of the many projects Edison was working on that year was an improvement to the telegraph. In 1876, he had invented a device that recorded the clicks of a telegraph transmission as embossed notches on a paper tape, so that the message could be sent at high speed and then slowly transcribed later. Then, in the summer of 1877, Edison found himself thinking about the work of a French inventor, Leon Scott de Martinville, who had built a device 20 years earlier that captured sound through a metal diaphragm and traced undulating lines, corresponding to the acoustic waves, on a metal drum. The Frenchman had never been able to play back the sounds he had traced, but Edison

BETTMANN-CORBIS

BROWN BROTHERS

Emile Berliner

The inventor, left, poses at age 75 with an early radio set. His flat disk and player showed up both Edison and Bell. He is also "Nipper's" father: the trademark for his invention showed a dog listening to "his master's voice" coming from a gramophone's speaker horn; it was later adapted by RCA

already had a playback mechanism. Substituting a metal diaphragm for the telegraph key, Edison bellowed the word "Hello" and watched the recorder scratch out a long, wavy indentation on the paper. When he ran the tape back through the player (this time substituting a telephone speaker for the telegraph), Edison later wrote, "we heard a distinct sound, which strong imagination might have translated into the original 'Hello.'"

Within a month, Edison improved upon his improvised gadget by sketching

BROWN BROTHERS

1902: EDISON CYLINDER

The 50¢ Gold Mould cylinder was obsolete when launched, beaten by Red Seal's $1 10-in. disk, which played longer—four minutes of music

1948: LONG-PLAYING RECORD
Hungarian Peter Goldmark, working for Columbia Records, invented the format. The LP's slow rotation—33 revolutions per minute—allowed much more content on each side of the disk. This scanning electron microscope image shows the stylus riding through the grooves, which rise and fall in proportion to the waves of the recorded sound

BROWN BROTHERS

COPY CATS
Unlike Edison's cylinders, Emile Berliner's disks could be stamped out in unlimited numbers and played back over and over. Above, making LPs in 1948

CORBIS

DR. TONY BRAIN—SPL—PHOTO RESEARCHERS

1949: 45 R.P.M. SINGLE
RCA's cheap, 7-in. disks held less than four minutes of music per side. Made of plastic, they allowed '50s teens to match their records with their hula hoops

out a device specifically intended to record sounds, substituting a metal drum wrapped in tin foil for the paper tape and a sharper needle for the blunt stylus on the telegraph recorder. His engineers took two months to build a working model; finally, on Nov. 20, Edison recited *Mary Had a Little Lamb* into the new device, then played it back. It was perhaps the only time in the career of the famed proponent of perspiration that a single burst of insight led directly to an invention.

Unfortunately, Edison's flash of brilliance didn't extend to what the phonograph (a word he coined for the new device) might be used for. He was mainly interested in bizarre schemes like making the Statue of Liberty talk and recording the dying words of famous people. Selling recorded music seemed like an afterthought, and even when he got around to turning his invention into a product, Edison offered a narrow choice of recordings that suited his personal taste.

Within a year, Alexander Graham Bell improved upon Edison's phonograph by replacing the tin-foil cylinder with one made of wax. Bell, who called his device a graphophone, could also replay his wax recordings many times, whereas Edison's could be played back only once. But in 1877, both Edison and Bell were trumped by Emile Berliner, a German inventor living in

1982: COMPACT DISC

The CD format improved sound quality (some audiophiles strongly disagree) and was smaller than the LP. By the end of the 1980s, CDs had supplanted their vinyl rival

STEVE PERCIVAL—SPL—PHOTO RESEARCHERS

Washington, who developed a flat disk to replace Bell's wax cylinder. The records for Berliner's Gramophone (made initially of glass, but eventually from a blend of clay and shellac) not only offered unlimited replays but could also be stamped out in unlimited numbers, while Edison's and Bell's cylinders had to be recorded one at a time.

For nearly a century, Berliner's disk would remain essentially the same—although its size, material and the speed at which it was played would evolve. The introduction of the LP, or long-playing record, in 1948 was the beginning of an avalanche in format changes that continues to make sound ever easier to record and play back. What has never gone away, though, is the wonder of listening to a voice, a noise or a note of music that first set the air pulsing in a time and place far from our ears. It was this original miracle that transformed an obscure tinkerer from New Jersey into the Wizard of Menlo Park. It's no surprise that when Edison was asked, late in life, to name the invention he was proudest of, he answered, without hesitation, "My phonograph." ■

2001: APPLE iPOD

This small wonder, the size of a deck of cards, plays digital computer MP3 files, providing good sound quality while holding up to 1,000 songs

Tape Recording

BETTMANN CORBIS

Valdemar Poulsen

While disks provide music for the masses, over the years many audio professionals and sound snobs have insisted on tape. Magnetic recording (rather than the mechanical kind pioneered by Edison) was invented in 1893 by Danish engineer Valdemar Poulsen, who was trying to build a telephone answering machine. Because signal amplification was still decades away, Poulsen's recordings (made on a steel wire) were almost inaudible, which caused him to abandon the technology.

By the 1930s, however, once radio pioneers had come up with a way to boost electronic signals, and wire had given way to flexible tape coated with metal oxides, Poulsen's invention had become not only practical but also in many ways superior to phonograph technology.

Poulsen's work led directly to the videotape recorder, which was invented in 1956 by Ampex Corp. researchers Charles P. Ginsburg, Charles E. Anderson and Ray Dolby—at the time a 19-year-old school dropout.

Dolby would become famous 10 years later when he devised a new kind of circuit that dampened the low-level signals that make up most audio interference, while leaving high-level signals untouched—dramatically improving sound quality in movies and on tape. Dolby went on to invent successor technologies, such as Dolby Digital Surround Sound, which is used in more than 25,000 theaters around the world. Ray Dolby, 70 in 2003, has received Oscar, Emmy and Grammy awards and the National Technology Medal.

AL SATTERWHITE

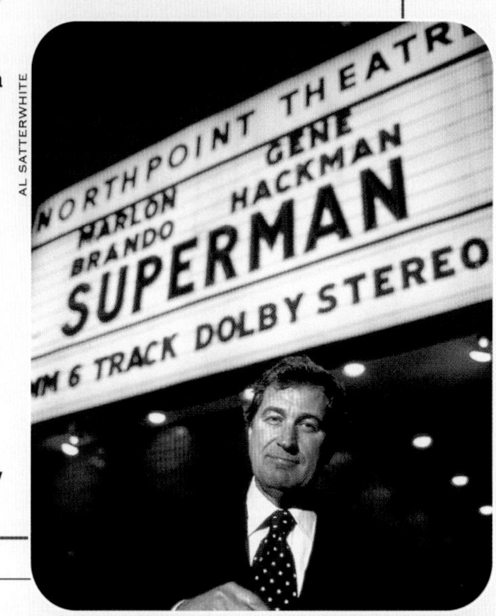

How We Eat

Cyrus McCormick and Early Mechanical Reaper

5

BROWN BROTHERS

1862: MILKING MACHINE

The first practical machine for milking cows was patented by Leighton O. Colvin and featured rubber cups that surrounded the cow's teats and created a vacuum to draw milk out. Colvin immediately sold the British rights to his invention for $5,000 but continued to improve upon the American version through the 1870s

Harvest of Change

How farmers, engineers, chemists and geneticists put dinner on the table

"The power of population is indefinitely greater than the power in the earth to produce subsistence for man," wrote economist and philosopher Thomas Malthus in 1798 in his *Essay on the Principle of Population.* "Population, when unchecked, increases in a geometrical ratio. Subsistence increases only in an arithmetical ratio." The population of the world when Malthus wrote those words was fewer than 1 billion people. Today there are more than six times that many souls covering the earth. So ... why aren't we all starving?

Consider this: in 1830, an American farmer needed 300 man-hours of labor and five acres of land to produce 100 bushels of wheat. By 1987, those same 100 bushels sprang from only three man-hours of labor and were grown on just three acres. In labor time alone, this is an increase in efficiency of thousands of percentage points.

What made the difference? In a word, technology. In the 1850s (when farmers made up more than half the U.S. work force), American growers started to replace the age-old energies used on the farm—human and animal muscle—with mechanical power. The impetus for change had begun in 1831, when a Virginia blacksmith's son, Cyrus H. McCormick, built the first practical mechanical reaper, a horse-drawn machine that harvested wheat.

The introduction of the two-horse straddle-row cultivator in 1856 more than doubled the amount of land a single farmer could manage. Of course, more land and more crops required more water. The Plains states have oceans of freshwater deep underground, but how to bring it up? The first self-governing windmill (a nifty machine that turned to face whichever direction the breeze was coming from, shut down automatically in high winds

1904: CATERPILLAR TRACTOR

The steam-powered traction engines Benjamin Holt designed for his farm machinery, right, were so heavy the equipment would often sink into the soil. So Holt replaced the wheels with a pair of linked treads 9 ft. long and 2 ft. wide. A photographer who shot the contraption said it looked a bit like a caterpillar, and the name stuck: Holt's company eventually became today's Caterpillar Corp.

WYOMING DIVISION OF CULTURAL RESOURCES

Barbed Wire

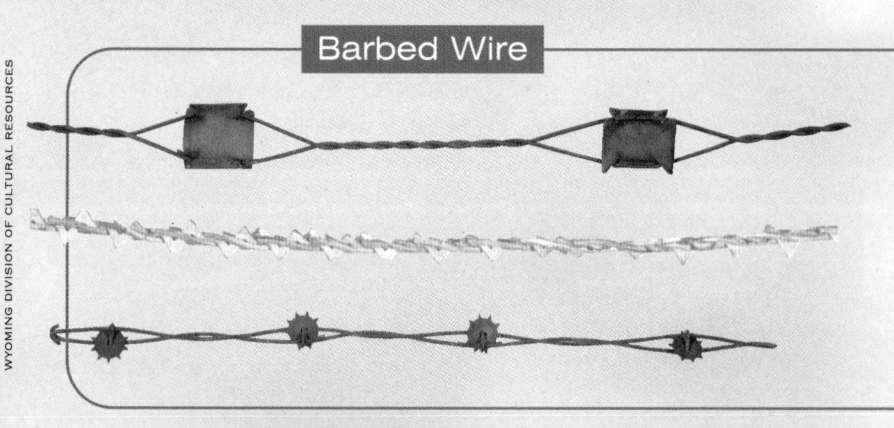

GO AHEAD: FENCE ME IN

Although barbed-wire fencing had been patented as early as 1868, Joseph Glidden perfected it with a design that locked the barbs in place and a machine that mass-produced it. His invention was the beginning of the end of the West's wide open spaces—and sparked brutal "range wars" between farmers and ranchers.

and pumped only as much water as a farmer needed) was invented by Daniel H. Halladay, a New England machinist, in 1854.

The bumper crop of innovation that followed reduced a farmer's work time, multiplied his strength and increased his acreage. It included the combine harvester (introduced in 1860), the combine seed drill (1867) and the sheaf-binding harvester (1878). By the mid-1870s, crop output had increased so dramatically that the first silos (for long-term storage of grain, which was now being grown faster than it could be used or sold) were erected. In 1892, an Iowa thresherman named John Froelich rigged one of the new gasoline engines to a mechanical harvester that had previously been drawn by horses, and he covered more acres in a few days than a team of livestock could in a month. Charles W. Hart and Charles H. Parr (who coined the word tractor) became the first to manufacture gas-powered farm machinery in 1901; six years later, Michigan auto-maker Henry Ford got into the business.

In 1862, L.O. Colvin patented the first reliable cow-milking machine. The increased production of both meat and dairy products was given a boost six years later, when William Davis, a Detroit fish merchant, invented the first refrigerated railroad car. Within a decade, perishables that had once been sold only in local markets were being shipped across the nation.

The dairy industry got two other

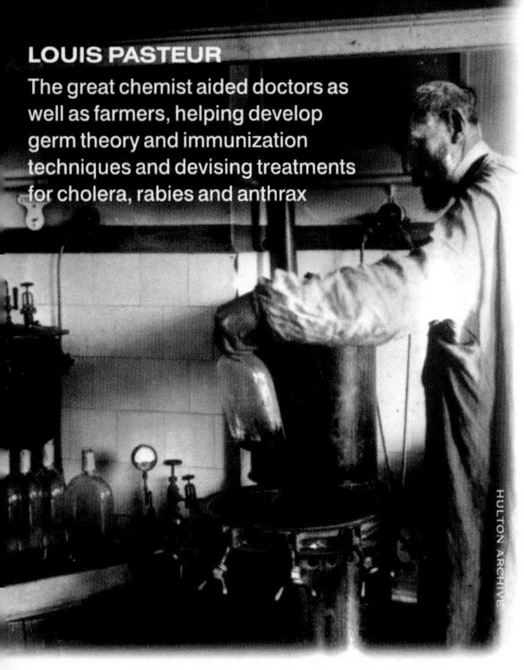
HULTON ARCHIVE

LOUIS PASTEUR
The great chemist aided doctors as well as farmers, helping develop germ theory and immunization techniques and devising treatments for cholera, rabies and anthrax

vital assists in the late 19th century, both from men who wore white coats rather than overalls. In 1864, French chemist Louis Pasteur realized the bacteria that caused wine to spoil could be rendered harmless by heating it to just below the boiling point. He soon applied the process to milk, and this process that kills harmful bacteria remains an essential part of sanitary dairy production.

Parisian engineer Auguste Gaulin patented the first mechanism for homogenizing milk in 1898. By pumping it through very small tubes under high pressure, he was able to reduce the fat globules to a size at which they would remain suspended within the milk, rather than separating and rising as cream. Among other advantages to his process: it made modern ice cream possible, just in time for the invention of the ice-cream cone six years later.

Farming's laboratory-born revolution peaked in the late 1950s, when plant biologists like Norman Borlaug created a "green revolution," using genetic science to create superior strains of existing crops—most famously, a dwarf wheat—that produced far greater crop yields on much less land. Borlaug's work is credited not only with saving hundreds of millions of lives in the Third World, but also with preserving hundreds of thousands of acres of wilderness from being converted to farmland.

Today farmers constitute less than 5% of the U.S. work force, but they are feeding more Americans (and more people around the world) than ever before. Thomas Malthus, you forgot to predict a Norman Borlaug. ■

ART RICKERBY—TIMEPIX

NORMAN BORLAUG
The Iowa-born agronomist, whose pioneering work saved both lives and land, was awarded the Nobel Peace Prize in 1970

HOW TO MAKE GOLDEN RICE

MARKUS BUHLER—LOOKAT PHOTOS FOR TIME

INGO POTRYKUS

Following in the footsteps of Norman Borlaug, German biologist Ingo Potrykus combined DNA from daffodils and bacteria with rice genes to cook up a revolutionary strain of rice that will provide a far more nutritious food for the malnourished. He calls it "golden rice"—critics denounce it as a "Frankenfood." Here's how it's made.

1 The genes that give golden rice its ability to make beta-carotene in its endosperm (the interior of the kernel) come from daffodils and a bacterium called *Erwinia uredovora*

2 These genes, along with promoters (segments of DNA that activate genes), are inserted into plasmids (small loops of DNA) that occur inside a species of bacterium known as *Agrobacterium tumefaciens*

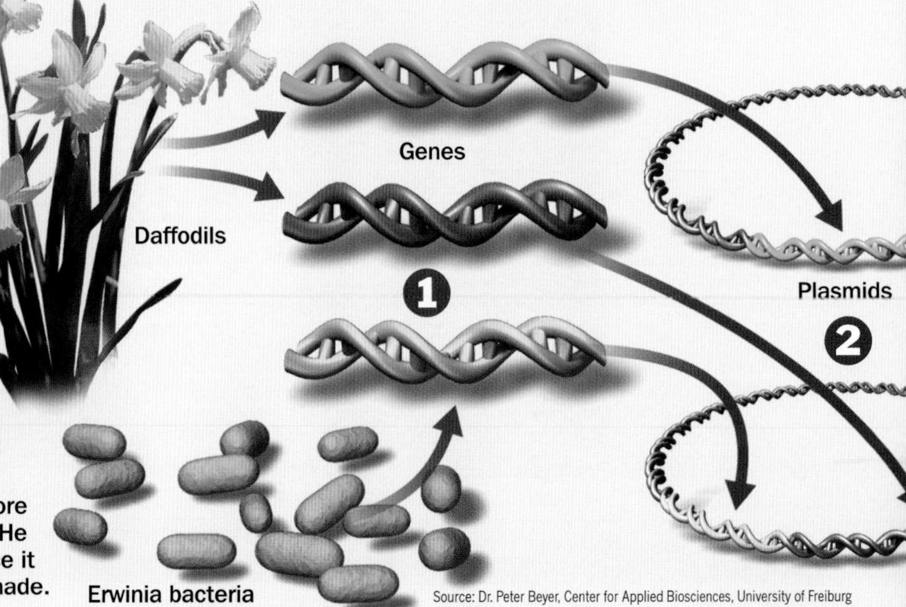

Source: Dr. Peter Beyer, Center for Applied Biosciences, University of Freiburg

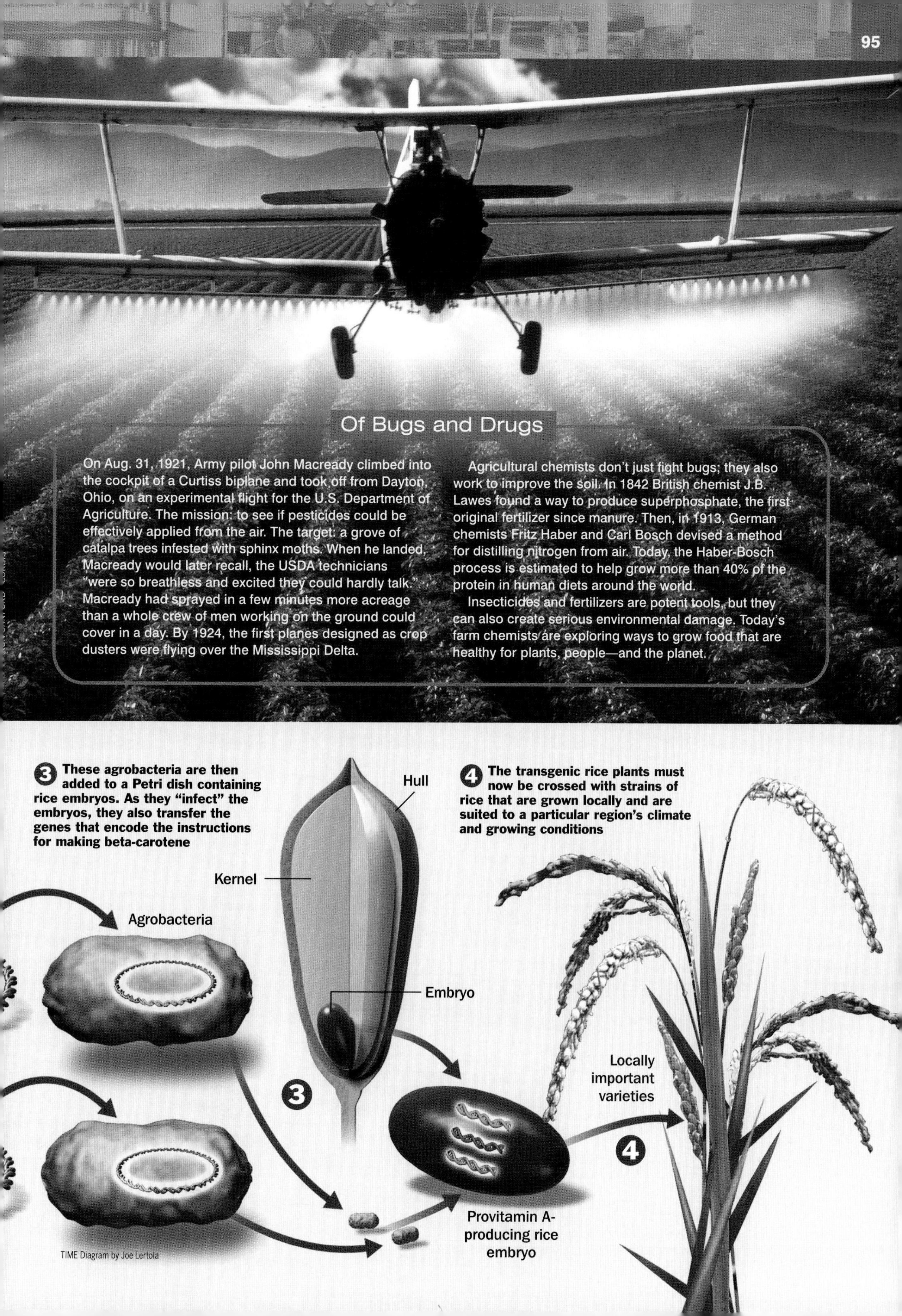

Of Bugs and Drugs

On Aug. 31, 1921, Army pilot John Macready climbed into the cockpit of a Curtiss biplane and took off from Dayton, Ohio, on an experimental flight for the U.S. Department of Agriculture. The mission: to see if pesticides could be effectively applied from the air. The target: a grove of catalpa trees infested with sphinx moths. When he landed, Macready would later recall, the USDA technicians "were so breathless and excited they could hardly talk." Macready had sprayed in a few minutes more acreage than a whole crew of men working on the ground could cover in a day. By 1924, the first planes designed as crop dusters were flying over the Mississippi Delta.

Agricultural chemists don't just fight bugs; they also work to improve the soil. In 1842 British chemist J.B. Lawes found a way to produce superphosphate, the first original fertilizer since manure. Then, in 1913, German chemists Fritz Haber and Carl Bosch devised a method for distilling nitrogen from air. Today, the Haber-Bosch process is estimated to help grow more than 40% of the protein in human diets around the world.

Insecticides and fertilizers are potent tools, but they can also create serious environmental damage. Today's farm chemists are exploring ways to grow food that are healthy for plants, people—and the planet.

TIME Diagram by Joe Lertola

BLOOM WHERE YOU'RE PLANTED

Carver turned down offers to earn more than $100,000 a year from Thomas Edison, Henry Ford, Mohandas Gandhi and Joseph Stalin. The last two wanted him to rebuild their nation's agriculture

Profile

Merlin of the Soil

With a peanut as his philosopher's stone, George Washington Carver transformed once-humble plants into high-paying crops

AS A CHILD OF BLACK SLAVES IN THE WANING days of the Civil War, the infant George Washington Carver was kidnapped from a farm in Diamond, Mo., by Confederate raiders and left to die in a roadside gully. He was recovered by a bounty hunter hired by his owner, Moses Carver, who regarded his slaves as family. Carver and his wife Susan nursed the baby back to health and decided to raise him as their own. They gave him the name George Washington Carver.

The ordeal left Carver frail for the rest of his life but also excused him from farming chores. He filled the time with constant examination of every plant he could find. Before he reached his teens, Carver had already earned a local reputation as "the plant doctor." At a time when schooling was all but unheard of for African Americans, he graduated high school and then became the first black ever admitted to Iowa State University.

> "He could have added fortune to fame, but... he found happiness and honor in being helpful to the world."

Carver's genius was unmistakable. In a few years, he joined the university faculty; in a few more, he was a nationally recognized expert in breeding, nurturing and caring for plants. This led to an invitation from black educator Booker T. Washington to start the botany program at the Tuskegee Institute in Alabama. It was here that Carver would do the work that secured his place in history.

The postbellum South was depleted and bled dry in more ways than one. Its legendarily rich soil was virtually ruined from centuries of producing cotton and tobacco, crops that drain the ground of nutrients. Carver was among the first to realize that while some crops defertilize the soil, others nourish it, drawing nitrogen—the essential ingredient in fertilizer—out of the air and depositing it in the ground. He proved that centuries of neglect could be undone in a single growing season. In 1897, after just one year of cultivating peanuts, soybeans and sweet potatoes, he was able to make depleted soil yield six times its usual harvest in crops like cotton and tobacco.

But harvesting valuable cash crops every second year, while recharging the soil for 12 months by growing produce for which there was little commercial demand, left farmers hard up. So Carver set about creating demand for the crops he had convinced the South to grow. His experiments with peanuts were revolutionary. By reducing this legume to its constituent chemical parts—fats, oils, gums, resins, pectins, sugars, starches and amino acids—then recombining them, he was able to create food products with more carbohydrate than potatoes, more vitamins than liver and nearly as much protein as beef.

Carver didn't actually invent peanut butter, but he did much to teach the American food industry that peanuts, once used mainly to feed pigs, had wide appeal for humans. Moving beyond food, he used peanut parts to create new iterations of products: bleaches, papers, plastics, shaving creams, synthetic rubbers, lubricants, dyes and cosmetics.

When he died in 1943, Carver's annual salary at the institute hadn't changed in 47 years: $1,500. With only three minor exceptions, he never tried to patent his work. He is buried on the campus; a simple stone above his grave bears the words: "He could have added fortune to fame, but caring for neither, he found happiness and honor in being helpful to the world." As a grave site, it sure beats a gully. ■

1928: PRESLICED BREAD

Once upon a time, bread was bought in loaves at the local bakery. Enter U.S. inventor Otto Frederick Rohwedder, who created the presliced-loaf and sealed-bag process in 1928. Two years later, Wonder Bread made presliced bread a rage. Next stop: the toaster. Briton Charles Strite invented the timed, pop-up toaster in 1919; Toastmaster introduced the beautiful appliance at left 10 years later

Cooking Up the New

The goals: fresher food, faster meals and leftovers that last longer

In the kitchen, where yesterday's tools are as welcome as last week's left-overs, innovation is constant. For the past 150 years or so, the emphasis has fallen on preserving food longer, cooking it faster and storing it more efficiently. In the preservation derby, the tin can has enjoyed a long run—it's almost 200 years old—but it remains *the* indispensable storage device. Rugged and recyclable, it seems set to carry on for some time.

The modern quest for convenience focused attention on the cooking process: How could the age-old oven be rethought and speeded up? The answer was discov-ered by accident, when Raytheon scientist

1818: TIN CAN

In 1809 French cook and inventor Nicolas Appert won a contest sponsored by Napoleon to create a container to keep food fresh; Appert's first efforts used glass containers sealed with wax, much like the Mason jars still used by home canners. Briton Peter Durand soon developed a tin-lined steel can to hold food, which he introduced to the U.S. in 1818. Almost 200 years later, the tin can still does its two jobs—protecting and preserving its contents—admirably. The good people at the Can Manufacturers Institute would like you to know that Americans use 200 million tin cans every day

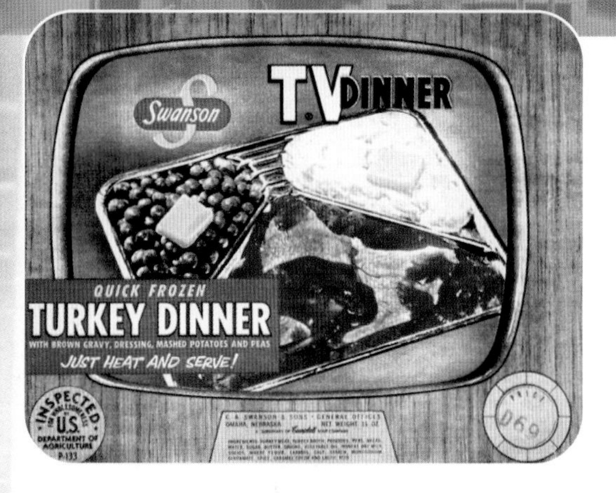

1904: THERMOS

Scottish physicist James Dewar's 1892 vacuum flask, below, had two layers of silvered glass with a vacuum between, keeping inside temperature steady. His student, Rheinhold Burger, improved the design by adding a screw-on rubber-seal cap

HAKE'S AMERICANA & COLLECTIBLES

1954: TV DINNER

More marketing triumph than technological breakthrough, this product paired a new notion—an entire frozen meal in a single serving dish—with a dazzling new technology, television. The A.C. Swanson Co.'s original menu reflects its supposed origin as the inspiration of an executive with an excess of Thanksgiving leftovers at home: turkey, stuffing and gravy, sweet potatoes and peas. Cost: 98¢

GRANGER COLLECTION

1913: STAINLESS STEEL

This metal hails from Sheffield, England's city of steel. Harry Brearley was hoping to create a metal that would resist heat inside a gun barrel; instead, after adding chromium to steel, he found a substance that resists corrosion. "Stainless" is the favored material for today's cutlery and appliances, as seen in the 1947 cruise-ship kitchen at left

1946: TEFLON

Answer to countless mid-scrub prayers, the nonstick plastic surface for cookingware was discovered by chance by DuPont chemist Roy Plunkett in 1938 and brought to market eight years later. Feeling scientific? Ask for polytetrafluoroethylene (PTFE)

BETTMANN CORBIS

1924: FROZEN FOOD

Inspired by a trip to Newfoundland, where he saw fish preserved in the freezing cold, Brooklyn-born naturalist Clarence Birdseye invented a quick-freezing process that retained the freshness of fish, vegetables and fruit

JESSICA WECKER

Percy Spencer was working with a magnetron, a tube that generates radio waves, and noticed that a candy bar in his pocket had melted. Experiments with (what else?) popcorn followed: Raytheon introduced its first, huge models in 1947, concentrating on the restaurant market.

If tin rules the front end of the cooking process, plastic has taken over the back end, in the form of wraps, pouches and polyethylene storage containers. And when it comes to chopping and blending, the kitchen may be a laboratory for serious engineers like Cuisinart guru Carl Sontheimer, but it is also a playground for old-fashioned gadget gurus like the irrepressible entrepreneur Ron Popeil and other equally dicey gizmo hawkers. ■

MICKY PFLEGER—TIMEPIX

1973: CUISINART

Born in New York City and raised in France, Carl Sontheimer, left, was an inventor and engineer who modified an industrial blender he first saw in France into a versatile chopper and blender

1976: POP-TOP CAN

Erma Fraze's 1965 pull-tab can aimed to make the can opener obsolete, but it left a trail of cut fingers, curses and cast-off tabs. Daniel Cudkzik's 1976 version, below, stays attached, foiling litterbugs

SERGIO DORANTES

1973: PLASTIC SODA BOTTLE

Engineers tried hundreds of plastics to replace glass soda bottles but were foiled by the carbonation in the drink, which exerts a pressure most plastics can't handle. Chemist Nathaniel Wyeth (of the artistic clan) found a new way to manufacture polyethylene that did the job

JAMES KEYSER—TIMEPIX

DART & KRAFT

1945: TUPPERWARE

Entrepreneur, former DuPont chemist and plastic pioneer Earl Tupper introduced his invention in 1945, just in time to hold all those leftovers made by the microwave oven, introduced only two years later. Tupper, who sold produce door-to-door as a New Hampshire farm boy, was also a marketing genius: to reach his audience of homebound postwar housewives, he introduced a social element into sales, inventing the "Tupperware Party," a model used by home marketers ever since

JAMES KEYSER—TIMEPIX

1953: SARAN WRAP

Discovered accidentally by Dow Chemical's Ralph Wiley in 1933, polyvinylidene chloride is a plastic barrier against water, acids, bases and solvents. It was first marketed to businesses in 1949 and to homes—as Saran Wrap—four years later

1947: MICROWAVE OVEN

A chef demonstrates the use of the first commercial microwave oven, built by Raytheon. The Radarange stood more than 5 ft. tall, weighed more than 750 lbs. and cost about $5,000. Each unit included plumbing hookups, since the magnetron tube generating the microwaves was water-cooled

ROY STEVENS—TIMEPIX

HULTON ARCHIVE

BROWN BROTHERS

1871: CHEWING GUM

People have been chewing tree resin and sweetened wax for centuries. But in 1871, U.S. rubber magnate Thomas Adams—with an assist from Mexican Antonio López de Santa Anna, conqueror of the Alamo, patented the first modern gum. It was based on chicle, a gummy substance found in Mexico's sapodilla plant. Bubble gum was created in 1906 and was brought to the market as "Blibber-Blubber"

1870: MARGARINE

Created by France's Hippolyte Mège-Mouries, who won a contest to create a substitute for butter, margarine threatened U.S. dairy farmers. Outlawed, colored, mislabeled, bootlegged, smuggled—it's the bad boy of the dinner table!

1941: M&M'S

Brainchild of master candy man Forrest Mars Sr., who also gave us the Mars Bar and the Milky Way, the chocolate disks originally were packaged in paper tubes. The "m" stamp on the shell appeared in 1950, to fight counterfeit candies

URBANO DEVALLE—TIMEPIX

Food of Thought

Who says there's no such thing as invented cuisine?

Few would argue that the popsicle ranks up there with the telephone as a great invention. Well ... few adults, anyway. But there is a purity to the Eureka moments behind our favorite treats—a cone-shaped waffle to turn ice cream into a finger food!—that makes them irresistible. Even their ad slogans sound like design specifications: "Melts in your mouth—not in your hand." Now, that's new and improved. ■

JAMES KEYSER—TIMEPIX

1894: COLD CEREAL

The flakes were flukes, discovered by accident in 1894 when W.K. Kellogg ran some stale wheat through rollers and baked the result *(see next story)*

1904: ICE CREAM CONE

St. Louis' most famous food—to everyone but serious Budweiser lovers—this magnificent creation was unveiled at the World's Fair celebrating the centennial of the Louisiana Purchase

BROWN BROTHERS

1840s: INSTANT GELATIN

The wiggly dessert was invented by 19th century polymath Peter Cooper, founder of New York City's Cooper Union school, and later marketed under the Jell-O brand name

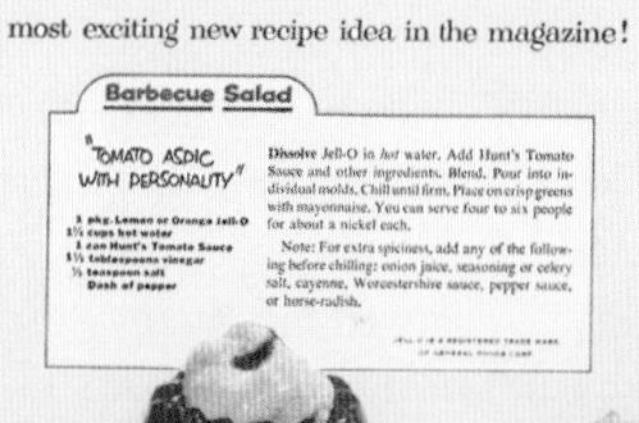

STEVEN FREEMAN

HAROLD J. TERHUNE—H&H PHOTO SERVICE

1937: SPAM

We know why this spiced ham product was a breakthrough: Who could imagine packaged meat that didn't spoil? We can guess why its name was borrowed for Internet junk mail. But no one can explain why it is the favorite food of Hawaiians

1886: COCA-COLA

Beloved as "the pause that refreshes" and reviled as mere sugar water, the brainchild of Atlanta pharmacist John S. Pemberton really did contain extracts of the coca plant in its early years. The magisterial bottle was adapted from a 1913 design by Earl R. Dean

1890: PEANUT BUTTER

Some provenance! Invented by nutrition maven Dr. John Kellogg, the gooey chewy made its bow (along with the ice cream cone) at the St. Louis World's Fair in 1904. It was later improved by George Washington Carver and first put on the market in 1922 under the brand name Skippy

PHILLIP JAMES CORWIN—CORBIS

1923: POPSICLE

This cool jewel was invented by—of course—an 11-year-old. In 1905, Frank Epperson of San Francisco left a glass of flavored soda water on the back porch with a mixing stick in it as the mercury dropped ... next morning: Voilà! He put the treats on the market as Ep-Sicles 18 years later, then sold his invention, which was wisely renamed the popsicle

TED THAI—TIMEPIX

Profile

Snap & Crackle's Pop

W.K. Kellogg peddled health, acquired wealth and changed the way we start the day—all by turning wheat into a morning treat

"EAT WHAT THE MONKEY EATS, SIMPLE FOOD AND not too much of it," advised Dr. John Harvey Kellogg, proprietor of Michigan's nationally famous Battle Creek Sanatorium. But the sanatorium's guests, late-19 century luminaries such as Henry Ford and John D. Rockefeller, didn't much care for monkey food. So Dr. Kellogg tasked his younger brother, Will Keith Kellogg, his assistant at the health resort, to come up with a grain-based health food.

At a time when most Americans' diets consisted of meat-and-potatoes or potatoes-and-meat, both Kelloggs were Seventh-Day Adventists, for whom meat was taboo. They were intrigued by the possibilities of cereal (named for Ceres, the Roman goddess of grain), first concocted in 1863 by another sanatorium operator, Dr. James Caleb Jackson, who mixed bran-filled wheat flour with water, baked it, then broke it into chunks and pieces. Following his lead, Will Kellogg spent months experimenting with wheat dough, boiling it, baking it, then fashioning it into sheets by feeding it through a rolling press.

> "Eat what the monkey eats, simple food and not too much of it."
> —Dr. John Kellogg

One evening in early 1894, Kellogg was preparing a batch of dough when he decided to knock off for the day. Returning in the morning, he discovered that his boiled dough had changed consistency; it was now brittle. When he pushed the mass between the rollers, crumbled flakes appeared—crunchy, toasty, tasty flakes. Served to sanatorium guests at breakfast the next day, Kellogg's find was a hit; second and third helpings followed, until the flakes ran out.

One of the guests was recuperating from nervous exhaustion following several failed business ventures. But when Charles W. Post tasted Kellogg's cereal, he sensed what his next pursuit would be. The Kellogg brothers quickly began selling their cereal; the recipe was patented—in Dr. Kellogg's name. The race was on. The recent guest quickly launched his own company, C.W. Post. Inventions breed rivalries; in this case, a duel that has lasted for more than a century began the day the innovation was introduced.

Will Kellogg believed passionately in his product and left the sanatorium to run the cereal business. But John regarded the flakes as a sideline. So W.K. Kellogg began covertly buying up the shares his brother had signed away. At the same time, he was cooking up a new recipe: corn flakes. Once W.K. Kellogg had acquired a majority of the stock in the Battle Creek Corn Flakes Co., he changed the name to Kellogg's. In 1914, he patented a new kind of wax-paper bag that would keep the cereal fresh inside its box (he called it Waxtite).

Within a few years, he had developed new recipes for incorporating bran into cereals (launching Bran Flakes in 1915 and All-Bran in 1916); Rice Krispies appeared in 1928. Kellogg proved equally adept at marketing. In 1907, he devised a brilliant promotion based on the slogan "Wednesday Is Wink Day," which promised homemakers a free box of cereal if they winked at their grocer on the appropriate day.

But this proponent of a balanced diet found little stability in his personal life. He never reconciled with the brother he deceived, and he forced his son John L. out of the family business in favor of John L.'s son John Jr.—then ousted John Jr., who later committed suicide. The man who reinvented the way American families start the day often ate breakfast alone. He continued to run Kellogg's until he was 85 years old, a lonely autocrat of the breakfast table. ■

SINGLE-MINDED
Kellogg, seen here in 1935, never stopped exploring new uses for familiar grains. His last creation, Kellogg's Special K, was introduced after he died

BROWN BROTHERS

1902: THE AUTOMAT
A commissary for a mechanized century, the Automat was based on a German model. It was brought to America by two young Philadelphians, Joseph Horn and Frank Hardart. Something about the slot-machine menu seems to stir us: the automat was celebrated in a 1930s song by Irving Berlin and was the subject of a 2002 book, *The Automat: The History, Recipes and Allure of Horn & Hardart's Masterpiece*

Fast, Faster, Fastest

For food's final frontier, look to outer space and the U.S. Army

There is little serious science or invention behind most fast food. But hard-driving Ray Kroc, the man who put a McDonald's into every sizable town in America, was a genius of marketing who took the measure of modern life and realized that the 20th century's faster pace demanded … faster food. Appropriating the assembly-line process that had revolutionized manufacturing and adopting the franchise plan that was beginning to remake retail marketing, Kroc was commerce's close equivalent to a great inventor, the visionary entrepreneur.

But Kroc followed the first marriage of haste and cuisine by a half-century. Horn and Hardart's 1902 Automat was a brilliant solution to the quest for a speedy feed, combining the variety of a cafeteria with the self-service convenience of a coin-operated vending machine. And as for *that* device, the first of these user-friendly (well, usually) gizmos appeared in London in the 1880s; it sold postcards.

Those pursuing real science in fast food are the researchers who are working to ensure that soldiers, astronauts and others can get a hot meal in their bellies, however far from a kitchen. ■

NASA—SPL—PHOTO RESEARCHERS

1961: SPACE FOOD
NASA entered the fast-food business when it began the Mercury program. Most of these packets of dried food required rehydration by a water gun

REUTERS—CORBIS

TIM BOYLE—GETTY IMAGES

1980s: MEALS READY-TO-EAT

A U.S. soldier in Afghanistan digs into a prefab feast. World War II soldiers chowed down on the canned Combat Individual (C) Ration. In the early 1980s the Pentagon upgraded to meals ready-to-eat, or MREs. Today's MRE specifications include a shelf-life of three years at 80°F, or six months at 100°F. Each meal contains at least 1,200 calories and is prepared by a flameless, water-activated, chemical heating element that is part of the unit. The tough pouch keeps the food fresh without dehydration or freeze-drying

1955: McDONALD'S

One of the first McDonald's, in Downey, Calif., is a time capsule of the 1950s, as photographed in 1990

GRANGER COLLECTION

1959: TANG

This dried orange-juice product was introduced to the general public in 1959 by General Foods. But it languished until the ad campaign for its1965 national roll-out made the drink synonymous with space flight. It is still frequently—and mistakenly—cited as an example of how NASA science has affected everyday life

GREAT INVENTIONS

How We Live

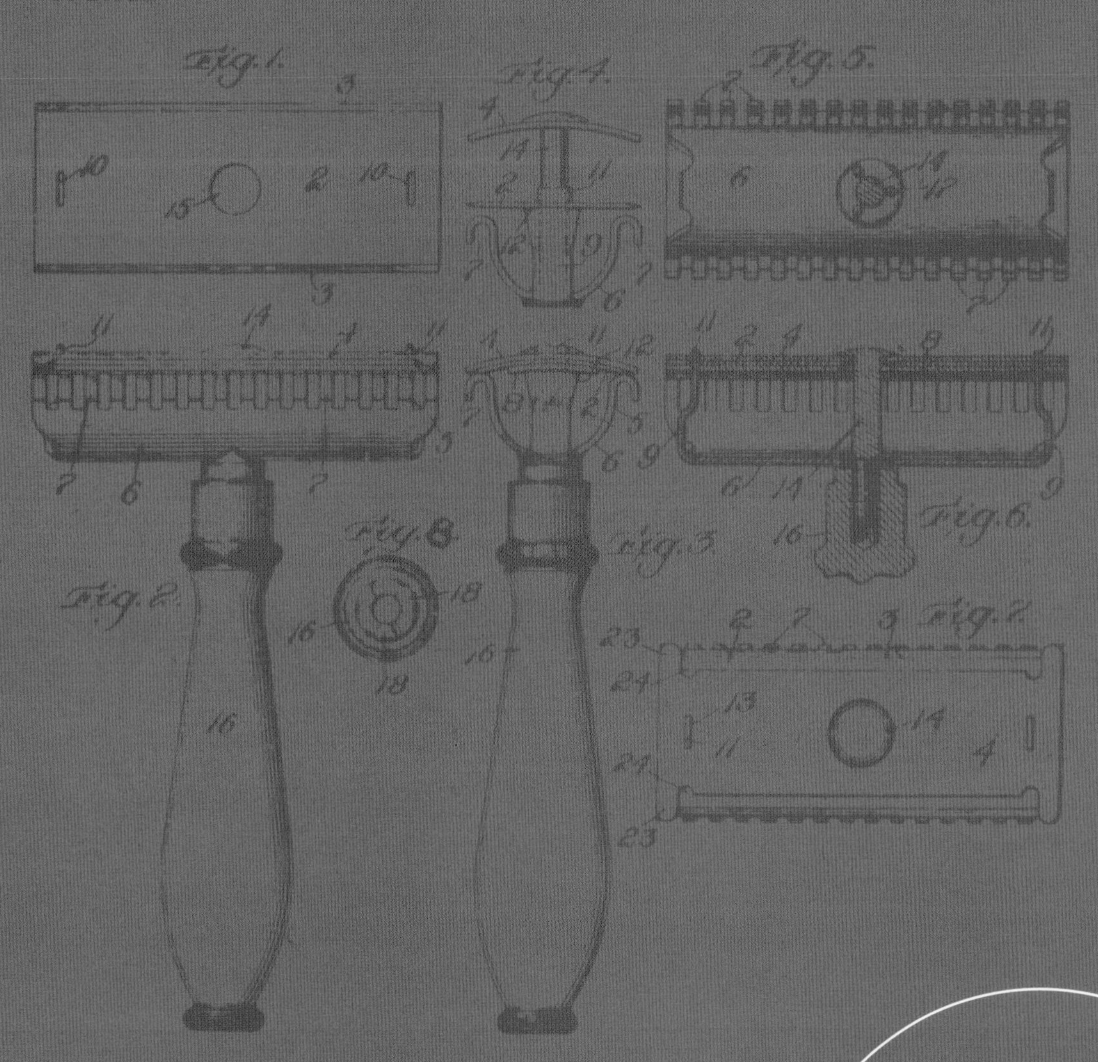

Singer Sewing Machine, circa 1876

Gillette Safety Razor, patent drawing, 1903

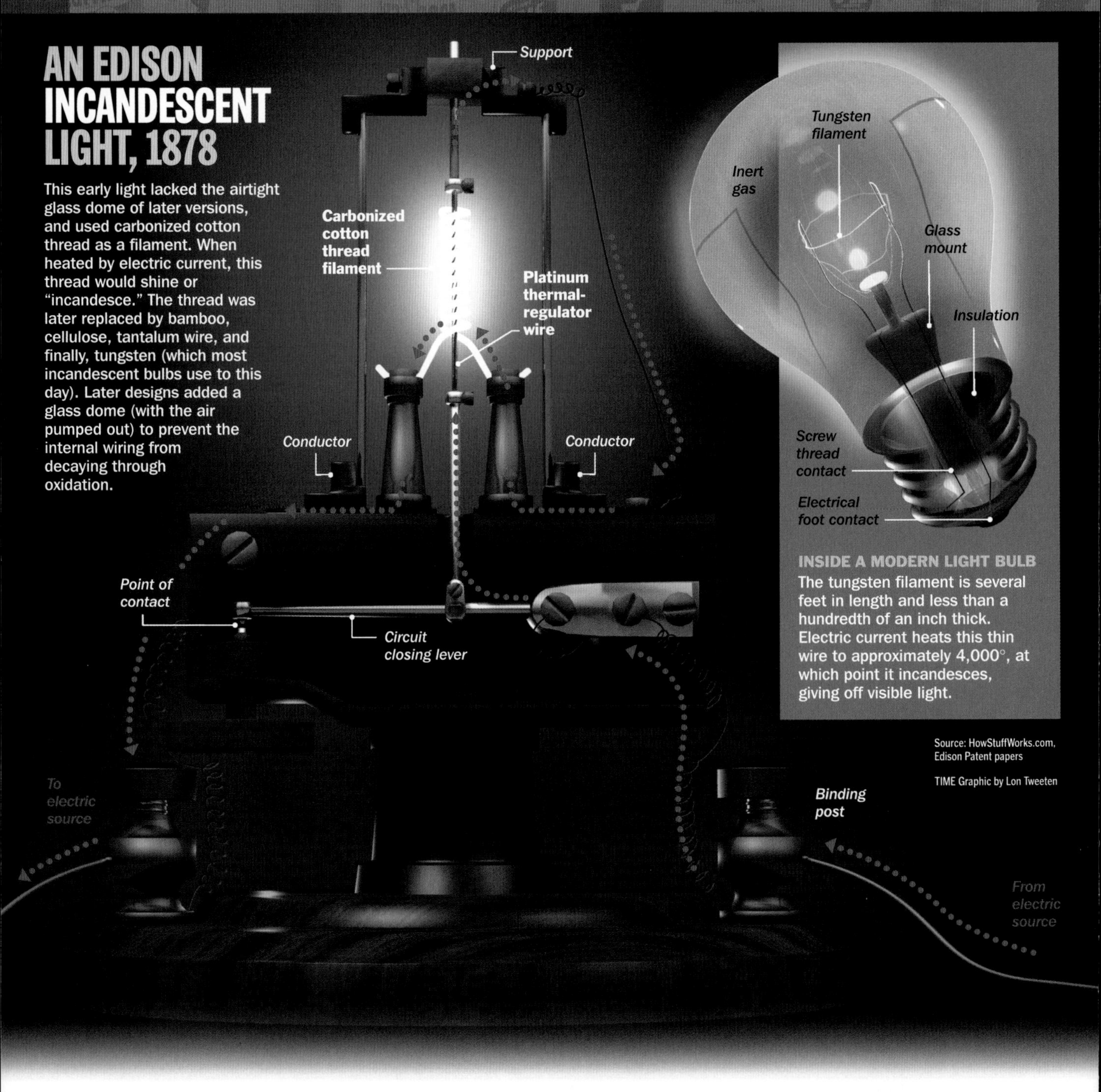

This Little Light of Mine

Thomas Edison lit up our nights—with an assist from the "Insomnia Squad"

Sorry, but Thomas Alva Edison did not invent the electric light bulb alone. That distinction is shared by a succession of engineers and tinkerers, probably beginning with English engineer Humphry Davy, who made a strip of charcoal glow by passing an electric current through it in 1809. For the next 70 years, a host of inventors came up with incremental improvements to what remained a laboratory novelty—changing the composition of the filament, housing it within a protective glass globe, reducing the air pressure inside the glass and so on. But every new light bulb suffered from the same "too" problems—it was too bright, too hot, too large, too fragile and too short-lived. As a result, although electric lights were sometimes used in large, outdoor spaces, they remained impractical for the home or workplace.

Edison knew there was a fortune to be made by the person who could find a way around these pitfalls. While he didn't begin with any unique or original insights about how to reach his goal, he did start with a method that no other inventor could match: the brute force exerted by a team of engineers willing to work around the clock and explore every promising path, test every possible variable.

Edison began by borrowing heavily from the bestl models he could find: bulbs patented in 1875 by Henry Woodward and Matthew Evans, and another designed in 1878 by British physicist Joseph Swan. That same year, Edison and his "Insomnia Squad" set out to systematically test more than 1,600 kinds of filament—everything from sewing thread to hair from a human beard. In 1879, Edison absent-mindedly pulled a bamboo shoot from a fan in his office and decided to try that. When the bamboo was baked into carbon and placed inside a glass bulb with all of the air evacuated, it shone for more than 40 hours—almost triple the life of any previous bulb.

GRANGER COLLECTION

1879: EDISON & CO. AT WORK
The "Invention Factory" team experiments with an early light bulb. When British physicist William Thomson was asked why no one else had improved the incandescent light, he replied, "No one else is Edison"

Deciding that bamboo was the best material, however, didn't end Edison's process: instead, it spurred him to send researchers scouring through South America and the Far East to find the best kind of bamboo. After experimenting with more than 1,000 different varieties, he settled on a Japanese import. Fourteen months after beginning his work with electric lights (and after purchasing Woodward and Evans' patent and taking Swan on as a minority partner), Edison crowed, "I have accomplished all that I promised." He deserved a place in the glowing flames of the limelight—which he'd just succeeded in making obsolete. ■

Variety Lights

FLUORESCENT

These lights dispense with a filament and instead use electric current to energize atoms of mercury, causing a phosphor coating on the inside of the bulb to glow. First designed by Nikola Tesla in the 1890s, they didn't become practical until the 1930s, when German inventor Edmund Germer perfected the mixture of mercury and inert gases inside the bulb. Result: a lamp that produces five times as much light as an incandescent bulb while consuming less power, producing less heat and lasting much longer.

CHRIS NIEDENTHAL—TIMEPIX

HALOGEN

Engineers at General Electric invented the first halogen lights in 1959. The bulbs, which heat pressurized halogen gas to more than 5,000 degrees inside a quartz chamber, shine far brighter than incandescent lights. They were originally intended for car headlights. GE quickly expanded their use to stadiums and shopping malls but didn't begin selling halogens for home use until the 1970s.

STONE—GETTY

WIZARD AND RUBE

Edison, in later years, shows off his first light bulb and a giant descendant. Henry Ford called him "the world's greatest inventor and worst businessman." Edison didn't know how to market his movies and recordings, and let financiers like J.P. Morgan wrest control of electric-power utilities from his grasp

Profile

Sweating the Details

Thomas Edison not only dreamed up the light bulb, the motion picture and the phonograph—he invented our idea of the inventor

HE WAS YOUNG TOM SAWYER AND ANCIENT Prometheus rolled into one—a hero from the heartland whose cornpone charm made the new, vaguely threatening rise of technology somehow palatable to 19th century Americans. Michigan-born Thomas Alva Edison—home-schooled, then a telegraph operator—patented his first invention, an electric vote-counting machine, at age 22. Within a year, he had built an improved stock ticker and sold the rights for $40,000. Then came an "electric pen" (it punched hundreds of tiny holes in paper while a writer traced letters, creating a stencil for countless duplicates) and a telegraph upgrade that could transmit four messages over a single line at once. Each gizmo earned him another small fortune. Before he was 30, Edison was rich. But he was still unknown.

Then, in 1877, the bright young man strolled into the offices of *Scientific American* magazine, slid an odd contraption across the editor's desk and invited him to turn the crank. Out of the box came a human voice that said, "Good morning! What do you think of the phonograph?" The editor was amazed; soon the world was. Edison became a celebrity. By the mid-1880s, the post office was delivering thousands of letters each week to his Menlo Park, N.J., laboratory, addressed simply to "Thomas Edison, U.S.A." or to "the Wizard of Menlo Park."

Along the way, Edison made his one purely scientific discovery, when he realized (while experimenting with light bulbs) that electric current could be made to jump across empty space from one conductor to another in a vacuum. Entranced but perplexed by the "Edison Effect," he abandoned his work on the phenomenon; it later became the basis of electronics.

Such detours were an inevitable part of a career that, while punctuated by spectacular successes, was also marked by frequent dead ends. Edison's scheme for distributing electricity via direct current would have required a power plant on nearly every street corner and thick cables using more copper than is known to exist in the world. And Edison struggled with personal demons: a relentless publicity hound, he forbade his staff of more than 500 assistants to take personal credit for their work. His family relationships were turbulent: he eventually payed his son Thomas Edison Jr. to change his name.

None of which detracts from his legacy. This untrained polymath (he had only a few months of formal schooling) harnessed his old-fashioned reasoning-by-analogy to a work ethic that drove him to experiment with literally thousands of variations of each invention. What Edison created, more than any single invention, was the idea of the inventor. Collaborating with a small army of assistants and experts, he created the first independent research laboratory, which he called his "invention factory." It earned him 1,093 patents, more than anyone else before or since. He is the father of modern research and development.

Edison spent the last years of his life trying, without success, to develop a new, domestic source of rubber from plants. (True to form, he tried 17,000 different possibilities before giving up.) He died at age 84 on Oct. 18, 1931, timing his exit perfectly: it was the 52nd anniversary of his invention of the incandescent bulb. At the insistence of President Herbert Hoover, the lights across America were momentarily dimmed in his honor. ■

Out of the box came a human voice that said, "Good morning! What do you think of the phonograph?"

BROWN BROTHERS

1851: SINGER
This is one of the first sewing machines, made only five years after Elias Howe was granted a patent for his stitching process. Isaac Singer paid Howe $15,000 for violating his patent

High-Tech Home Ec.

Machines enter the house as inventors take the labor (O.K., a lot of the labor) out of sewing, washing and cleaning

When a French inventor, Barthelemy Thimonnier, introduced a primitive sewing machine in 1830, Parisian tailors replayed the storming of the Bastille, destroying his plant and all 80 of his machines. Theirs was a minority view: soon after its introduction, the sewing machine revolutionized Victorian-era home life, saving countless hours of difficult, close work while improving the quality and durability of clothing.

Although many tinkerers tried to build a mechanized stitcher, the first one to come close was a Massachusetts Quaker, Walter Hunt, who also did pioneering work on rifles and safety pins. But Hunt feared his machine might put tailors out of work, so he abandoned it.

A fellow citizen of the Bay State, Elias Howe, is generally credited as the father of the modern sewing machine. A former worker in a textile factory, he became a chronic invalid and was unable to work, so his wife took in sewing to support the family. Watching her stitch, Howe decided to craft a sewing machine. His design copied a

BROWN BROTHERS

Evolution of the Washing Machine

The Laundromat concept was pioneered in 1851, during the California Gold Rush; the machines used donkey power. But whether you're washing by hand, by donkey or by feeding quarters into the big monster down at the Laundromat for an extra-large item—it's all about the cycles. First, wash in soapy water. Second, rinse in fresh water. Third, wring it all out.

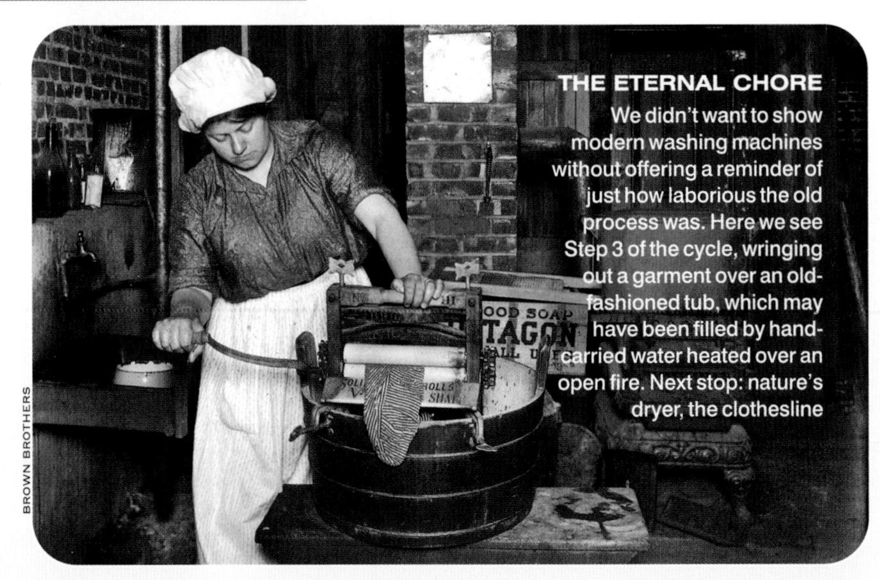

THE ETERNAL CHORE
We didn't want to show modern washing machines without offering a reminder of just how laborious the old process was. Here we see Step 3 of the cycle, wringing out a garment over an old-fashioned tub, which may have been filled by hand-carried water heated over an open fire. Next stop: nature's dryer, the clothesline

BROWN BROTHERS

key breakthrough from Hunt's 1833 machine: it used two spools of thread and a needle with an eye at the point. When the needle pushed through the cloth, it created a loop of thread on the far side; a shuttle would then slip thread through the loop, creating a tight lock stitch.

But the public was not quick to embrace Howe's machine, even though it outstitched five seamstresses in an 1845 demonstration. He went to England, then capital of the world textile industry, in hopes of selling his machine there, but a series of swindles and poor business decisions left him impoverished. When he returned to America, he found that his device was now wildly popular, its design hijacked and improved by competitors like Isaac Singer, a consummate marketer but not an inventor. Howe's misery only deepened when his wife died.

After years of court battles, Howe's 1846 patents were upheld, and he became a multimillionaire—as did Singer. Worn out from his battles, Elias Howe died in 1867, at only 48. His patent expired the same year. ■

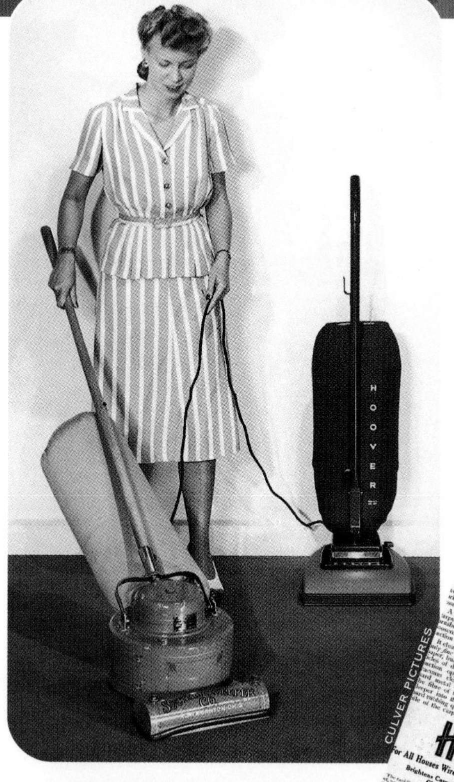

CULVER PICTURES

1907: HOOVER MODEL O

James Murray Spangler, a janitor in an Ohio department store, thought his brush-style carpet sweeper was causing his allergic reactions. He put together a contraption that used an electric fan motor to create suction and a pillowcase inside a soap box in which to catch dust. He stapled them to a broom handle—end of story. His cousin loved hers, and her husband, William Hoover, soon ran the company. Above, a 1950s publicity shot shows the very first model alongside a then current machine

1940S: DISHWASHER

Josephine Cochran invented a hand-cranked dishwashing machine in 1889, supposedly after complaining, "If no one else will invent one, I will!" The first electric machines appeared in the early 1900s. Below, a late-1940s top-loaded model

GORDON COSTER—TIMEPIX

C. 1910: WASHER

This early machine was made two years after the Hurley Co. introduced the first electric washer, invented by Alva Fisher.The wringer is still hand-cranked. Water often spilled over the sides of early machines, creating a serious hazard when it splashed onto the electric motor or switch

SCIENCE MUSEUM—TOPHAM-HIP—THE IMAGE WORKS

1946: MAYTAG WASHER

This advertisement sings the praises of Maytag's top-of-the-line model. The ad copy stressed the machine's "gyration action" (we call it agitation) and its ability to keep water hotter. But Step 3, the wring-out cycle, is still performed by hand

1947: WASHER

With its expanded cabinet, this machine looks more like today's washers. But its real innovation is inside: a "spin cycle," developed by the Bendix Corp. in the 1930s, whirls the clothes around, and the centrifugal motion wrings them out

MARTHA HOLMES—TIMEPIX

1924: CENTRIFUGAL AIR CONDITIONER
Carrier installed this historic machine in a J.L. Hudson & Co. department store. The first centrifugal air conditioner, it incorporated a new rotary engine that made it far more efficient than the up-and-down piston design of earlier models

CULVER PICTURES

Chill Out, Daddy Cool

Willis Carrier invented air conditioning to fix a problem in a printing plant

Some inventions, such as radio's Audion tube, arrive in this world amid a flurry of paternity suits, as inventors litigate over who made what connection first. But there is no question as to who's the father of air conditioning: the title belongs to Willis Carrier, a master engineer who cooked up his cool-down machine at age 21, only a year after graduating from Cornell University.

BROWN BROTHERS

Carrier was working for the Buffalo Forge Co. in 1902—designing, of all things, heating systems to dry lumber and coffee—when he was asked to study a problem at the Sackett-Wilhelms Lithographing & Publishing Co. in Brooklyn, N.Y. The high temperature generated by its big printing presses was combining with muggy summer air to affect the paper during print runs: pages wouldn't stay aligned properly, so the four colors used in the process were off-register, yielding a muddy mess rather than sharp illustrations.

Carrier's solution was to channel the air inside the plant through a system that condensed a refrigerant liquid—that is, one that evaporates at an extremely low temperature—thus absorbing heat and humidity in the process. A few years later, in 1906, he installed a larger system at a South

Carolina cotton mill. The same year he took out a patent for his "Apparatus for Treating Air"—the first of 80 acquired by this inveterate tinkerer.

Initially, Carrier's machines found only industrial applications. But in the 1920s, department store, theater and hotel owners realized the appeal of his process. When air conditioning first hit movie houses in the 1920s, the response made it clear that if your marquee didn't boast, "It's Cool Inside" ... well, your business might not be so hot.

Carrier later created a centrifugal machine that pumped air in circular fashion, turbine-style, a more efficient process than the reciprocating-piston design it replaced. He brought out his first home model, the Weathermaker, in 1928. Like his other designs, it used chlorofluorocarbons as a coolant. It wasn't until the 1970s that scientists realized these substances damage the earth's ozone layer. Today's air conditioners use environmentally sound refrigerants—and Carrier's company was a leader in adopting the new standards. ■

RON JETER—ANTIQUE FAN COLLECTORS ASSOCIATION

EMERSON ELECTRIC FAN

Fans have been powered by servants, by wind-up motors, even by oil. The first electric fan went on sale in 1882; this model, sporting six brass blades rather than four, was a luxury item in its day

BALDWIN H. WARD & KATHERYN C. WARD—CORBIS

From Ice-box to Fridge

Take away the little engine on the top of the early refrigerator at left, and you'd have an old-fashioned ice-box. Icemen delivered large cubes of ice to these units every few days, for they were not particularly efficient at retaining cool air. Refrigerators work on the same principles as air conditioning, using refrigerant liquids to cool down the air inside.

1923: SEEGER

The Seeger refrigerator, above, has been fitted with a General Electric motor that cycles the air inside past coils filled with a refrigerant liquid, which undergoes a constant process of compression and expansion, removing heat from the air as it expands

CULVER PICTURES

1940s: REFRIGERATOR

The machine at right still resembles the wooden model above, with the exposed motor blowing hot air away from the cooling coils. But its all-metal body and its more efficient internal insulation help it retain cool air far more efficiently than its predecessors

SCIENCE MUSEUM—TOPHAM-HIP—THE IMAGE WORKS

1935: SHELVADOR

Manufactured by Crosley, USA, this is one of those machines whose model name bespeaks its appeal: it is the first refrigerator to include a nifty design advance, the use of the space on the inside of the door for a series of shelves. In addition, the entire unit has a more sleek appearance, since the motor and exhaust fans have now been included within the main body of the machine

1917: SNEAKER

The first canvas tops were heat-sealed onto rubber bottoms by rubber magnate Charles Goodyear, who invented the vulcanization process in the 1890s. Keds were the first mass-market sneakers to use Goodyear's process. Until the advent of Nike in 1972, most pro athletes wore Adidas, left, or Converse

Ready to Wear

When technology plays designer, clothes get zippier

If we agree with Shakespeare that "the apparel oft proclaims the man," what do the trends in dress over the past 150 years proclaim about us? Clearly, that we also agree with architect Ludwig Mies van der Rohe: less is more. As with another necessity, food, recent clothing innovations stress convenience and simplicity, abetted by new fabrics and nifty closure devices.

So farewell, corset; hello, bra. Goodbye (at last!) buttonhook; hello, zipper. Farewell, garter belt; hello, pantyhose. Goodbye, leather shoes; hello, canvas-topped, gel-filled, light-emitting-diode-sporting, sports-icon-signed sneakers. On second thought, maybe less isn't always more. ■

CULVER PICTURES

From Corset to Bra

In 1913 New York City debutante Mary Phelps Jacob rejected the heavy whalebone corset (like the one at left) laid out for her before a formal dance: it was uncomfortable, and it poked through her sheer gown. So Jacob lashed two handkerchiefs together with a pink ribbon, wrapped this get-up around her bust, and went corsetless. After many of her friends clamored for similar attire, Jacob (under the pseudonym "Caresse Crosby") patented her new "Backless Brassiere" (from the French word for bodice) in 1914. In the 1930s, Ida Rosenthal and Enid Bissett's Maidenform bra was the first to categorize women by the "cup size" of their busts.

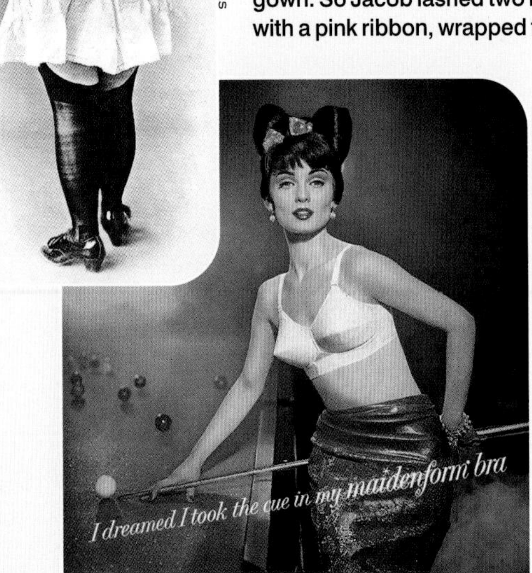

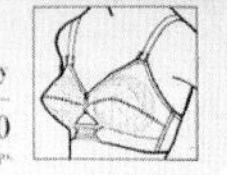

1893: ZIPPER

Now *that's* a great invention. Chicago inventor Whitcomb Judson called his 1893 triumph a "clasp locker." The word zipper was coined by the B.F. Goodrich Co. to describe the sound Judson's invention made when used to seal their galoshes, which featured zippers starting in 1923

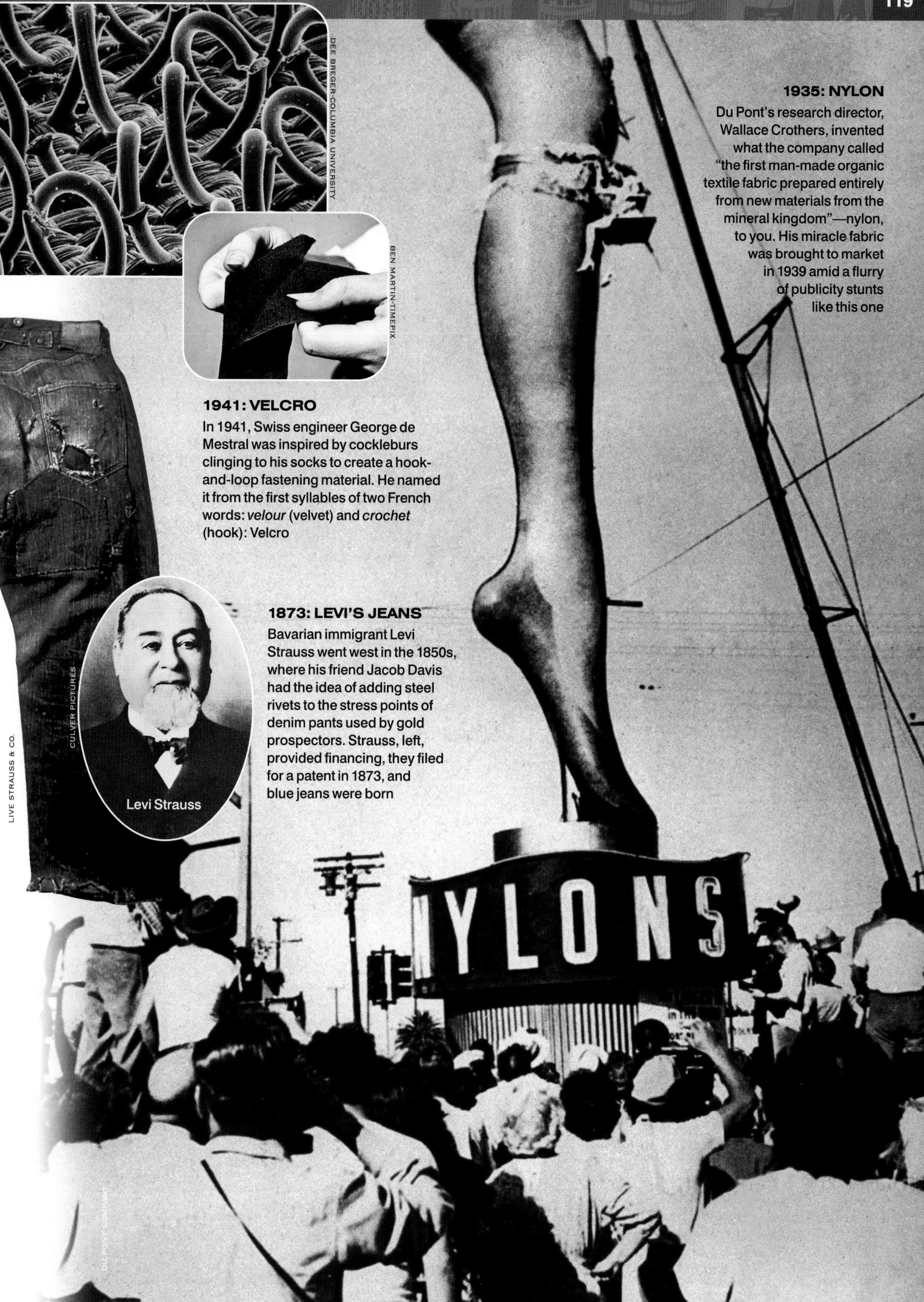

DEE BREGER-COLUMBIA UNIVERSITY

BEN MARTIN-TIMEPIX

1935: NYLON

Du Pont's research director, Wallace Crothers, invented what the company called "the first man-made organic textile fabric prepared entirely from new materials from the mineral kingdom"—nylon, to you. His miracle fabric was brought to market in 1939 amid a flurry of publicity stunts like this one

1941: VELCRO

In 1941, Swiss engineer George de Mestral was inspired by cockleburs clinging to his socks to create a hook-and-loop fastening material. He named it from the first syllables of two French words: *velour* (velvet) and *crochet* (hook): Velcro

1873: LEVI'S JEANS

Bavarian immigrant Levi Strauss went west in the 1850s, where his friend Jacob Davis had the idea of adding steel rivets to the stress points of denim pants used by gold prospectors. Strauss, left, provided financing, they filed for a patent in 1873, and blue jeans were born

Levi Strauss

CULVER PICTURES

LIVE STRAUSS & CO.

DU PONT COMPANY

Up Close and Personal

Well, somebody had to invent all that stuff that clutters up your cabinet

In the main, we can thank lucky accidents and talented amateurs for the innovative products for personal hygiene and grooming that crowd our medicine cabinets. In a 2002 poll, Americans voted the toothbrush their all-time favorite invention (it beat the zipper and some other serious contenders). Yet British political prisoner William Addis, who invented the modern toothbrush in 1770, remains unsung. (By his day, toothpaste had been around for centuries—Hippocrates' favorite formula involved three burnt mice and a hare's head.)

Band-Aids and Q-Tips were crafted by men who had no experience with

1892:TOOTHPASTE TUBE

Toothpaste is age-old, but dental hygiene wasn't well served by the packaging: a communal vat into which the entire household dipped their brushes. In 1892, a Connecticut physician, Washington W. Sheffield, patented the collapsible metal tube that allowed individual—and sanitary—portions for each member of the family

WATERPROOF
BAND-AID
WHITE

A SPEED BANDAGE FOR MINOR INJURIES

Johnson & Johnson
NEW BRUNSWICK N.J. CHICAGO, ILL.

JAMES KEYSER-TIMEPIX

1921: BAND-AIDS

Johnson & Johnson employee Earle Dickson was exasperated that his wife always seemed to be cutting herself in the kitchen. So he designed a small, sterile-gauze pad mounted on surgical tape that anyone with a small wound could apply without assistance. Cost: 98¢

GASLIGHT ARCHIVE

Give Him an "Extra Bathroom" for Christmas!

Christmas Special

NEW 1940

SCHICK DRY SHAVER

equipped with the new patented "WHISK-IT"

All Schick Dry Shaver Models operate on AC or DC

FREE! ...with all Christmas Schick Dry Shaver models, an ingenious new improvement, the "Whisk-it". Fitted to the shearing head, it catches all whisker clippings, removes last vestige of fuss and muss from shaving. Takes shaving out of the bathroom. Permits shaving anywhere, in any room, at home or office, before or after dressing!

He can shave after dressing, if he likes, as he reads his morning newspaper, or looks over his morning mail.

Now he can shave anywhere in the house; any electric outlet becomes an extra bathroom, so far as shaving is concerned.

CULVER PICTURES

1903: SAFETY RAZOR

In 1895, traveling salesman King Gillette was shaving when lightning struck: a blade "made cheap enough to do away with honing and stropping and permit the user to replace dull blades by new ones" would be worth a fortune. It took until 1903 to perfect the product, but in a year he sold more than 100,000 of his "safety razors." Canadian Jacob Schick patented the first electric model in 1928

ROYALTY FREE-CORBIS

CULVER PICTURES

1923: Q-TIPS

Polish immigrant Leo Gerstenzang decided that he could no longer abide his wife's cleaning their infant daughter's ears by wrapping damp cotton around the end of a toothpick and jabbing the tiny spear into the baby's auditory canal. Leo improved upon Mrs. Gerstenzang's design by attaching sterile cotton swabs to flexible cardboard sticks

1915: LIPSTICK

Legend credits Cleopatra with having invented lip color. The next improvement took a while: U.S. inventor Maurice Levy designed a slide-and-twist metal dispenser that went mainstream. In 1950 American chemist Hazel Bishop cooked up—in her kitchen laboratory—the first smudge-resistant, all-day, "kiss-proof" lipstick

THOM LANG-CORBIS

medicine. The safety razor was the brainchild of a traveling salesman who wanted to get rich by inventing a product that people would use once or twice and then throw away. (King Gillette succeeded at creating such a device, but never did get rich—he was forced out of the company he founded.) Lipstick was perfected by a woman who hoped to go to medical school but couldn't because of the Great Depression. And the first deodorant was cooked up by a man who didn't go to medical school but claimed that he had.

Salute this collection of characters next time you regard yourself in the mirror and find that your teeth sparkle, your skin is tan and your freshly shaved cheek feels as smooth as … well, as the baby's bottom below. ■

HULTON ARCHIVE-GETTY

1936: SUNTAN LOTION
L'Oréal founder Eugene Schueller made the first effective sunscreen, Ambre Solaire, in 1936. In 1944 Miami beach pharmacist Benjamin Green mixed cocoa butter and jasmine to aid sunburned U.S. G.I.s in the South Pacific. He named his goo "Coppertone"

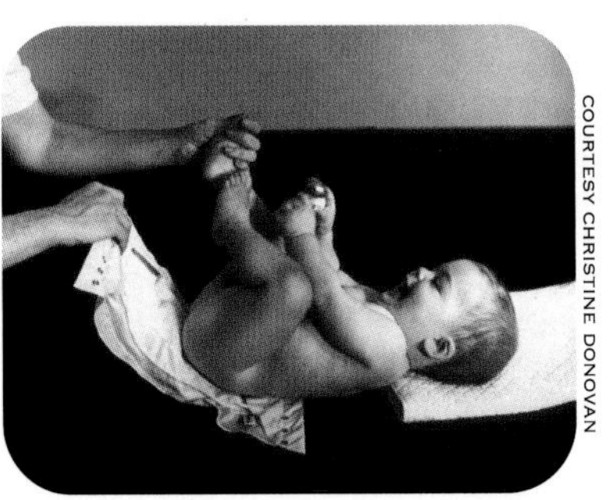

COURTESY CHRISTINE DONOVAN

1951: DISPOSABLE DIAPERS
New York City homemaker Mary Donovan cut up her shower curtain and made a waterproof diaper with an absorbent lining fastened with snaps in 1948. She sold the rights for $1 million in 1951. Procter & Gamble introduced disposable Pampers 10 years later

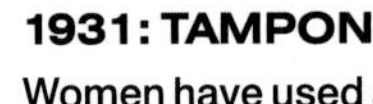

PHOTODISC GREEN-GETTY

1931: TAMPON
Women have used absorbent cloth to stanch menstrual flow since at least the time of the Pharaohs. The first modern tampon with an applicator was invented by Colorado physician Dr. Earle Haas in 1931. He sold the rights to his "catamenial device" (the name came from the Greek word for "monthly") two years later for $32,000

PHOTODISC BLUE-GETTY

GASLIGHT ARCHIVE

1903: DEODORANT
Mum, the first deodorant, was invented by a Philadelphia patent-medicine man who chose to remain anonymous to avoid legal trouble. Mum's formula contained zinc oxide (which killed some of the bacteria that cause body odor). It was applied by hand to the underarms

1888: CONTACT LENS
Swiss physician Eugen Fick devised the first practical contact lens, lubricating his version (a thick, blown-glass dome that covered the entire eye) with sugar water and grape juice. In 1962, Otto Wichterle and Drahoslave Lim developed the first usable "soft" contact lenses, made of flexible, porous plastic

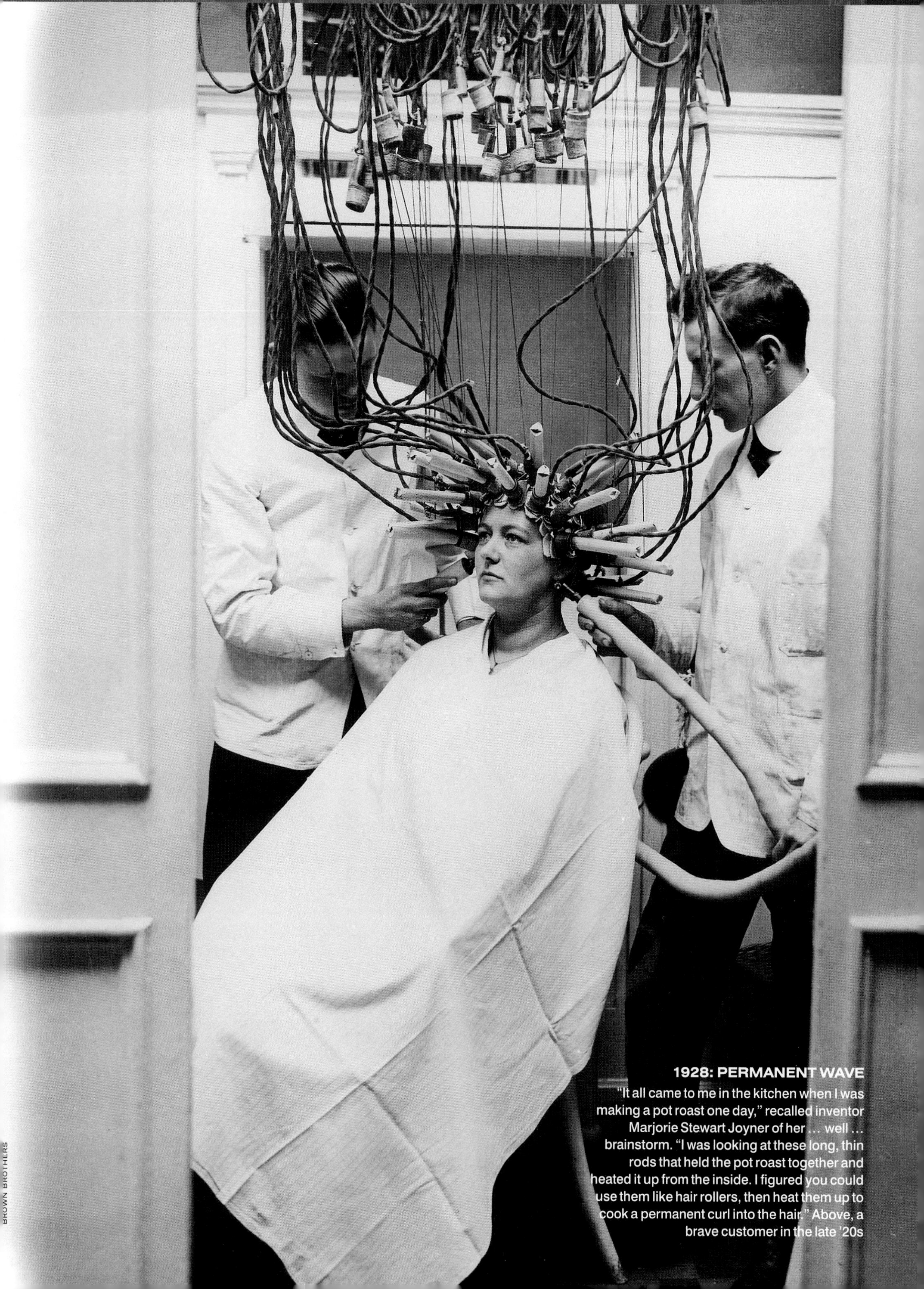

1928: PERMANENT WAVE

"It all came to me in the kitchen when I was making a pot roast one day," recalled inventor Marjorie Stewart Joyner of her ... well ... brainstorm. "I was looking at these long, thin rods that held the pot roast together and heated it up from the inside. I figured you could use them like hair rollers, then heat them up to cook a permanent curl into the hair." Above, a brave customer in the late '20s

BROWN BROTHERS

All Around the House

Now, if they'd just invent a pull-cord that actually started a lawn mower ...

"Country gentlemen will find in using my machine an amusing, useful and healthful exercise," British engineer Edwin Budding said of his 1830 invention, the first practical lawn mower. Budding's contraption was improved upon in 1899 by inventor John Albert Burr, who patented a mower featuring clog-resistant rotary blades and traction wheels. By 1961, gardening historian Charles B. Mills would write of gasoline mowers that "if all of them in a single neighborhood were started at once, the racket would be heard 'round the world." ■

1919: GAS-POWERED LAWN MOWER

In 1919, retired U.S. Army Colonel Edwin George took a motor out of an old washing machine, mounted it on a lawn mower and called the result a Moto-Mower. It was the first lawn mower to run on gas

WALTER SMITH/TAXI/GETTY

1930: PLASTIC TAPE

3M chemist Richard Drew invented transparent tape shortly after the advent of cellophane. 3M's Scotch brand tape supposedly got its name from an early test, during which an irate customer told Drew, "Take this tape back to those Scotch bosses of yours and have them put more adhesive on it!"

CORBIS

SCIENCE MUSEUM, LONDON-TOPHAM-HIP-THE IMAGE WORKS

1861: YALE LOCK

Linus Yale Jr., who inherited his father's locksmith shop, invented the first cylinder pin-tumbler lock, in which a small, flat metal key with a serrated edge is slipped into a hole too small for anything other than a pin. Each lock has a unique "combination" configured into the edge of the key

1943: AEROSOL SPRAY CAN

During World War II, more G.I.s were dying in the South Pacific from bug-borne diseases than from actual combat. Desperate for a new way to spray insecticides, the Pentagon called upon researchers Lyle Goodhue and William Sullivan, who invented aerosol spray powered by liquefied hydrocarbon gases. The first propellants turned out to damage the earth's ozone layer; today's models are planet-friendly

BROWNIE HARRIS-CORBIS

1957: BUBBLE WRAP

Shredded newspapers were the standard packing material for everything from lead weights to ancient Greek vases until engineers Al Fielding and Marc Chavannes (who were trying to perfect plastic wallpaper) invented Bubble Wrap in a Hawthorne, N.J., garage

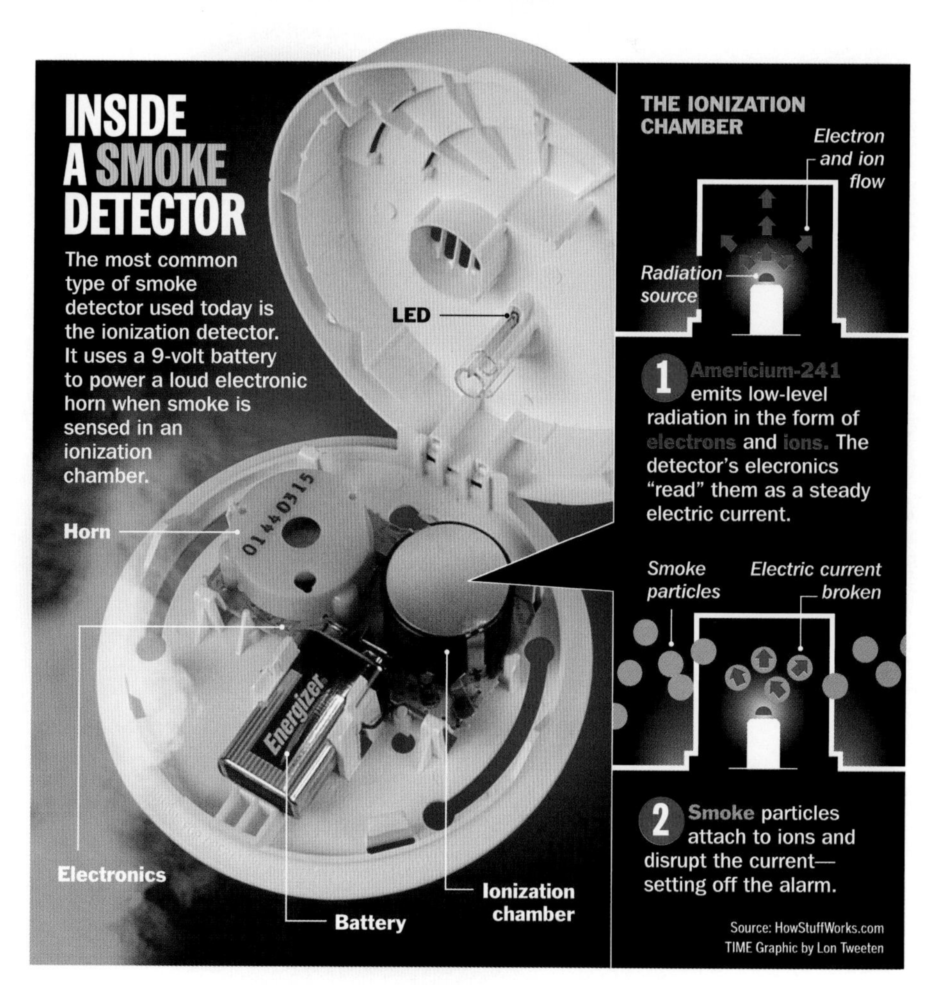

How We Think

George B. Grant Calculating Machine, 1876

7

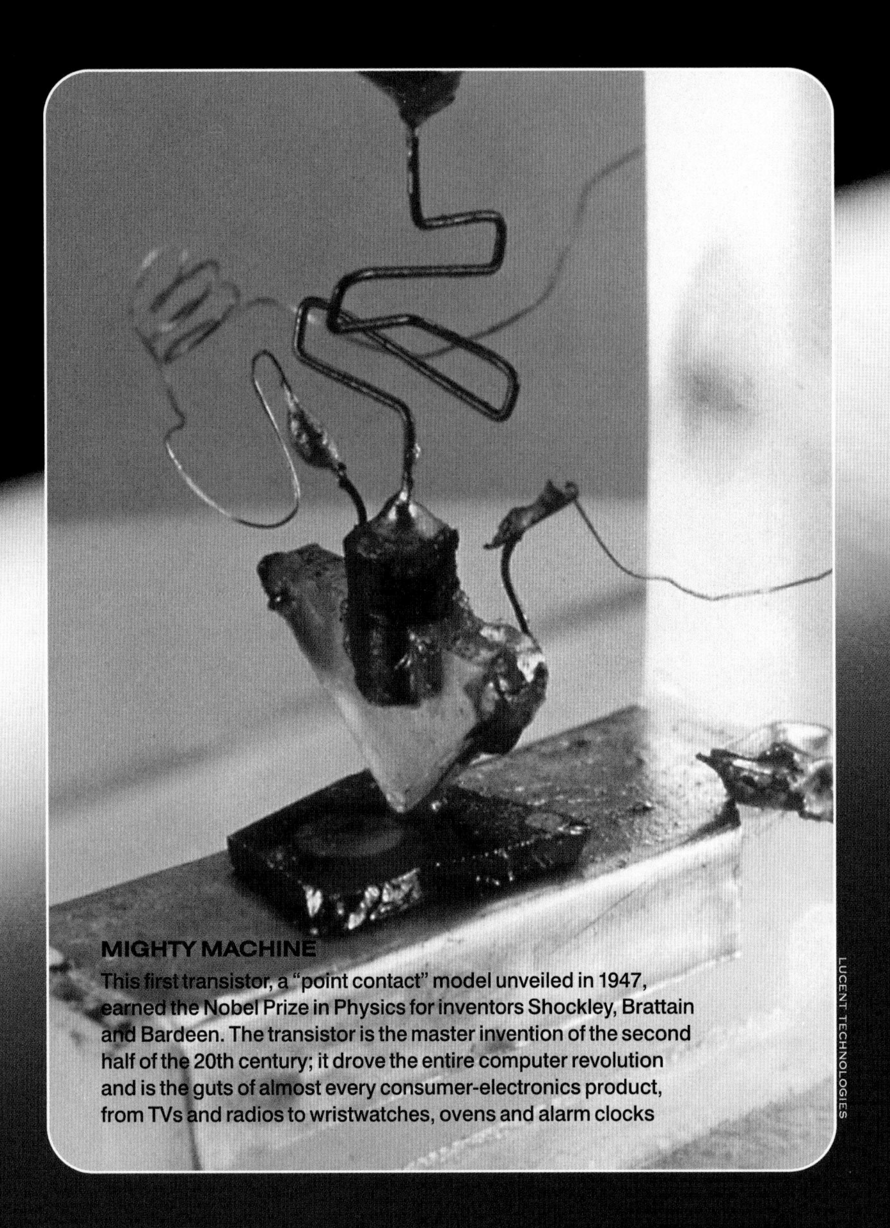

MIGHTY MACHINE
This first transistor, a "point contact" model unveiled in 1947, earned the Nobel Prize in Physics for inventors Shockley, Brattain and Bardeen. The transistor is the master invention of the second half of the 20th century; it drove the entire computer revolution and is the guts of almost every consumer-electronics product, from TVs and radios to wristwatches, ovens and alarm clocks

LUCENT TECHNOLOGIES

Downsized Dynamo

The transistor and its offspring powered a revolution that has yet to end

Computers are, at the bottom of it all, nothing more than an elaborate batch of switches, which reduce their processing to the binary language of 1s and 0s, registered as electrical pulses. The scientists who built the first computers used electronic vacuum tubes similar to those used in radios as switches. But they soon ran up against three barriers. The tubes were large: each one took up several cubic inches in machines that needed tens of thousands of tubes. They were expensive: up to $1,000 each—and each machine used thousands. And they generated staggering levels of heat that caused frequent burnouts.

Enter William Shockley, Walter Brattain and John Bardeen, scientists at Bell Labs who created the first transistor in December 1947. Using a class of materials known as semi-conductors (because, under certain predictable conditions, they conduct electric current, while in other predictable cases, they block it), Shockley, Brattain and Bardeen were able to rig up a circuit that would sometimes transfer electricity and sometimes resist it—hence the name.

Even the earliest transistors were a fraction of the size of vacuum tubes, used far less power and created almost no heat. So computer designers were able to build faster, cheaper, smaller machines. By 1958, transistors had

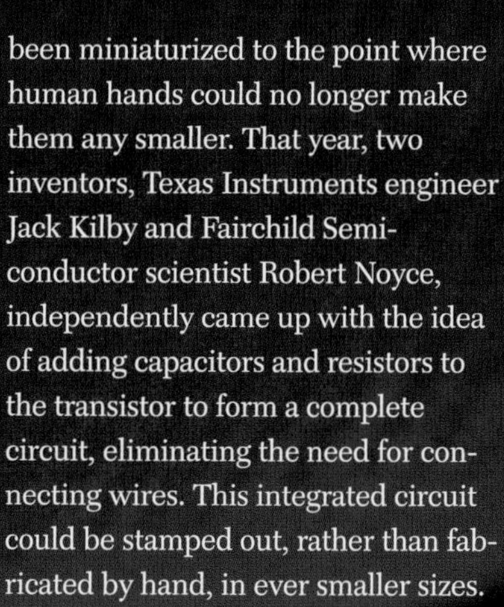

1971: MICROPROCESSOR
Intel's first tiny chip, the 4004, left, contained 2,300 transistors. Today the company can cram more than 77 million circuits onto a silicon wafer of similar size and sell it for about the same price

been miniaturized to the point where human hands could no longer make them any smaller. That year, two inventors, Texas Instruments engineer Jack Kilby and Fairchild Semiconductor scientist Robert Noyce, independently came up with the idea of adding capacitors and resistors to the transistor to form a complete circuit, eliminating the need for connecting wires. This integrated circuit could be stamped out, rather than fabricated by hand, in ever smaller sizes.

In 1965 computer scientist Gordon Moore was astounded to realize that an integrated circuit that had cost $1,000 in 1959 had fallen in price to $10. This inspired him to formulate Moore's Law: that the number of transistors on a chip would double every 12 to 18 months, reducing the price of each new generation of computers while increasing their speed.

STEPHEN GROHE—CORBIS

SILICON WONDERS
Microchips are made of silicon wafers, each containing hundreds of processors. Silicon is the preferred material because it is an inexpensive semiconductor that resists current leakage

Moore teamed up with Noyce to found Intel in 1968. Three years later, Intel engineer Ted Hoff launched the next revolution by adding memory and programmability to integrated circuits. His new invention was called the microprocessor. Intel's first version, the 4004 chip, measured just one-eighth by one-sixteenth of an inch yet contained 2,300 transistors—and it had more computational power than the 1946 ENIAC computer, which occupied an entire large room.

On the horizon, the possibility of "atomic computing"—in which a single electron is conveyed over a filament just one atom in diameter—beckons. If researchers succeed in current experiments with this theoretical nanotechnology, computer chips may increase in speed and power, while decreasing in size, at a pace that might make even a Gordon Moore dizzy. ■

1946: ENIAC
Although the Electronic Numerical Integrator and Calculator weighed 30 tons, took up 1,800 sq. ft. and used 18,000 vacuum tubes, it had less computational power than a modern digital watch. Designers John Mauchly and J. Presper Eckert Jr., inset, had "bugs" in their system: the warmth of all those vacuum tubes attracted moths

HULTON ARCHIVE—GETTY

Shrinking the Colossus

Why think big? The computer made its largest impact by getting smaller

For centuries, human beings have built mechanical contraptions to assist mathematical calculation. Ancient Babylon had the abacus; Renaissance Europe had the Jacquard loom, which used punch cards to remember complex patterns for tapestry weaving. Yet all these devices were mechanical one-trick ponies, and inventors dreamed of building a single machine that could be adapted to perform all sorts of tasks.

In 1820, British inventor Charles Babbage designed (but was unable to build) a mechanical "Analytical Engine" that would use punch cards like those of a Jacquard loom to solve math problems. Some card-sorting tabulators were built in the late 1800s, but Babbage's vision wasn't realized until the first half of the 20th century, when the mighty workhorse of electronics, the vacuum tube, allowed scientists to substitute electrical current for Babbage's mechanical gears and switches.

In the late 1930s, a brilliant Harvard graduate student, Howard Aiken, talked International Business Machines president Thomas Watson into bankrolling his Automatic Sequence-Controlled Calculator, the Harvard Mark I, which could solve complex differential equations. But it was a highly specialized machine, and as the prospect of war became clearer, military planners were demanding more versatile machines that could crunch numbers at higher speeds.

Vannevar Bush

HULTON-DEUTSCH COLLECTION—CORBIS

British mathematician Alan Turing was the first to build a truly programmable calculating machine, called Colossus. Completed in December 1943, it took up eight separate rooms, had almost no memory and couldn't store its programs. Instead, it was set up each day by an army of technicians who "programmed" it by manipulating hundreds of plugs and switches. Still, Colossus could perform in two hours calculations that would take human beings eight weeks to finish—and it went on to crack the Germans' "Enigma" code

While Turing was creating Colossus, U.S. mathematicians John Mauchly and J. Presper Eckert Jr. were building an Electronic Numerical Integrator

Computer Prophets

It's one thing to see the future; it's another thing entirely to see it, then help build it. Three of the fathers of modern computers did both. Alan Turing, then a 24-year-old British mathematician, published an obscure technical paper in 1936 that envisioned a machine that could solve any math problem. Within a decade, he had built the first such device, Colossus, and had used it to crack Germany's "unbreakable" wartime codes.

Alan Turing

U.S. scientist Vannevar Bush helped kick-start the computing revolution while at M.I.T. in the 1930s,with his Differential Analyzer, a mechanical system that solved high-level equations. In 1945 he wrote an article for the *Atlantic Monthly,* "As We May Think," that described his vision of what he called the "memex," or memory extender. This imaginary device consisted of a desktop equipped with a typewriter-like keyboard and a viewing screen that would act as a calculator, word processor, picture editor, mailbox and electronic filing cabinet. The first personal computers, so aptly described, eventually appeared—32 years later.

In 1946 Hungarian-born polymath John von Neumann wrote a report on the future of computers that called for them to store programs (rather than having to be reset manually before each use) and be able to run many different programs. A computer he later built, at Princeton's Institute for Advanced Studies, was the first to incorporate both these features.

John von Neumann

1943: COLOSSUS

British WRENS hand-program Turing's machine. The British government denied it existed until 1970

SCIENCE & SOCIETY PICTURE LIBRARY

and Computer—ENIAC, for short. Intended primarily to calculate trajectory tables for artillery crews, it didn't begin service until a few months after the war ended. But with 18,000 vacuum tubes to the Colossus' 1,500, it represented a huge advance in speed and computational power.

Mauchly and Eckert's innovations began changing the world of business in 1951, with the introduction of Remington Rand's UNIVAC (Universal Automatic Computer), which was faster, cheaper and smaller than all previous computers. The UNIVAC was followed in 1953 by IBM's 701 model, soon used by research labs, aircraft companies and government agencies.

1982: MS-DOS
Bill Gates bought the rights to the operating system, then built it into the standard of the computer world

HULTON ARCHIVE—GETTY

The computer era was booting up quickly, thanks to the rapid revolution in processing power driven by the transistor, the integrated circuit and the microprocessor. IBM's 7000 series introduced the transistorized computer to businesses in 1959. By this time, IBM was the dominant maker of the large machines known as mainframes.

Even as computers were growing in power, a new generation of inventors set out to make them smaller and easier to use. In 1977 computers took a giant step forward with the launch of the Apple II personal computer, the brainchild of two young "hackers," Steven Wozniak and Steve Jobs.

Just as Henry Ford aimed to put a car in every garage, Apple's founders hoped to put a computer on every desk, even at home. Their first model wasn't up to that goal, but when they introduced a new version, the Apple Macintosh, in 1984, its operating system incorporated advances—a new graphical interface, or desktop, and a handheld "mouse"—that finally made computers easy for the layman to run

Though late to the party, IBM and other makers of large computers soon rushed their own personal computers, or PCs, to market. IBM's PC worked on MS-DOS, an operating system licensed from Microsoft Corp. Given IBM's existing clout, and with copycat "clones" flooding the market, MS-DOS became the standard operating system, and Microsoft's hard-driving boss, Bill Gates, became synonymous with the computer age.

Microsoft's stranglehold on the computer's innards has been threatened in recent years by the Linux operating system designed by an idealistic young Finn, Linus Torvalds, who first offered his system for free download over the Internet in 1994. Linux is chipping away at the Microsoft hegemony, aided by competitors who resent its tough policies (after a long investigation, the company reached a deal with the U.S. government in 2001 to halt certain anti-competitive practices).

1977: APPLE II
Early Apples didn't come with a monitor; instead, they plugged into a TV screen

APPLE

Today, the dream of Charles Babbage has been achieved, and the Apple Computer dream has been semi-achieved (there is indeed a computer on almost every desk—it's just not an Apple). There is little doubt that this master machine, easily among history's most significant inventions, will keep getting smarter, smaller and friendlier—almost as if it had a mind of its own. ■

1973: CRAY 1 SUPERCOMPUTER
These giant machines that synchronize the work of several computers, like this one at Lawrence Livermore National Laboratory in 1981, were once jokingly described by their designer, Seymour Cray, as "mostly plumbing" because of the exotic liquid-cooling systems needed to keep densely packed microcircuits from overheating

Alan Kay

Brilliant computer researcher Alan Kay was recruited by Xerox Corp. in the late 1960s to head up development at its Palo Alto Research Facility (PARC). There, Kay and his team built the Alto, a desktop computer that featured overlapping windows, icons and a point-and-click environment that perfected the ideas for a graphical interface that had been pioneered by Douglas Engelbart, founder of the Stanford University Research Institute.

Xerox never brought the Alto to market, but Apple Computer founders Steve Jobs and Steven Wozniak took a good long look at it during a visit to PARC and drew on many of Kay's innovations when building their revolutionary Macintosh in 1984—the same year they made Kay an Apple Fellow, free to pursue his interest in the future of computers. Right, Kay with an early Macintosh.

CAROL A. FOOTE

WHO'S "THE SUIT"?
They're in the right places—Wozniak at the controls of an Apple II, Jobs supervising—in this 1979 shot. But the sight of Ur-geek "Woz" in a suit and tie must have been a shock to his hacker buddies

Profile

Homeward Bound

Computers? They were big monsters with big price tags, owned by big companies. Until Steven Wozniak and Steve Jobs came along

"HELLO? THIS IS HENRY KISSINGER, AND I'D like to speak to the Pope." Two geeky boys in their late teens were testing out the first gadget they had built together—a "blue box" that tricked AT&T's computer network into allowing free long-distance telephone calls. So they dialed the Vatican and asked to speak to the Pontiff, hanging up only when a Bishop came on the line to act as translator.

Steve Jobs and Steven Wozniak had attended the same high school in Los Altos, Calif., outside San Francisco, but they truly bonded as members of a cabal of computer hackers and techies, the Homebrew Computer Club, after each dropped out of college in the early 1970s. They toiled at low-level jobs in what would later be called Silicon Valley: Jobs at Atari and "Woz" at Hewlett-Packard. Jobs, a visionary entrepreneur, believed that people would pay handsomely for a new appliance that no large company had the vision or the nerve to sell—a personal computer. By 1975, Woz, the design man, had cobbled together some off-the-shelf circuit boards and connected them to a TV monitor. They had a machine. Now all they needed was a company.

To finance their start-up, Jobs sold his Volkswagen bus and Woz hocked his prized scientific calculator. Recalling a summer job in an Oregon orchard, Jobs christened the company Apple. The partners, operating out of his parents' garage, managed to sell a few hundred of their primitive machines (price: $666). In 1977, they developed the Apple II, which featured color graphics (when hooked up to the customer's TV set), smart packaging, and the ability to run a variety of software programs. The Apple II had annual sales of $139 million by 1980, and the company went public. The two twentysomething geeks were now each worth hundreds of millions of dollars.

The most remarkable feature of the Apple II was its unremarkability. It offered no new technology but combined existing components to create a radically new product. Competitors pounced: in 1981 IBM bought the hardware and software for its own personal computer, the PC, from outside vendors, slapped the Big Blue logo on the front and soon was far outpacing the original personal computer in sales.

But Apple kept its design lead. In 1984, the company launched the Macintosh, the first personal computer to feature a mouse, graphic icons, onscreen windows and pull-down menus. In a stroke, the arcane text-based systems used by all other PCs seemed about as modern as a Model T.

The most remarkable feature of the Apple II was its unremarkability. It offered no new technology

Again, most of the Mac's nifty accessories had been developed years earlier, but Apple was the first to get them to consumers. And again, Apple's innovations inspired another company (Microsoft) to develop a copycat product (Windows) for IBM machines that was so successful it trounced the company that started it all.

The unassuming Woz left Apple for good in 1985; the charismatic Jobs has had an on-again, off-again relationship with the firm. Since 1997, he has been back at its helm—and Apple is innovating again. 1998 brought the sleek new transparent iMac (which returned Apple to profitability). 2001 brought the nifty portable iPod MP3 player, topped off in 2003 with the breakthrough iTunes—the first online music service embraced by the record industry. Apple once again is the sodbuster of Silicon Valley. And by the way: Guess what computers Vatican City favors? ■

ROBERT FOOTHORAP

GREAT INVENTIONS

How We Work

Bessemer Steel Converter, illustration, circa 1860s

8

BRIGHT LIGHTS, BIG CITY
In 1880 Thomas Edison arranged the first practical demonstration of streetlights, illuminating the intersection of Broadway and 34th Street in New York City, today's Herald Square

Power Struggle

Alternating or direct current? When the lights went on, the battle began

Perfecting the incandescent light bulb might have caught the world's attention, but Thomas Edison knew he wouldn't have a business until he could generate and distribute the electric current needed to power his bulbs. Four years after Edison accomplished the former, he saw to the latter. By 1882, he had set up the world's first electric utility on Pearl Street in lower Manhattan, to supply electric power for 59 paying customers. Coal-fired, it was to be the model for a system that would send power coursing into every home and office in America. Or so Edison thought.

Edison's system for distributing power had little of the elegance of his other inventions. He favored direct current, which could be generated and transmitted at only a single voltage, which meant that any appliance needing more electricity would not work, and any designed to run on less would burn out. What's more, direct current could be conveyed over wires only for half a mile or so, meaning large, coal-burning power stations would soon be nearly as common as bus stops.

Enter George Westinghouse, who had recently invented a system for transmitting natural gas efficiently over long distances using very high pressure, then reducing that pressure to more moderate levels for household use. He suspected that a similar approach could work with electricity. Nikola Tesla—a former protégé of Edison's, now his rival and Westing-

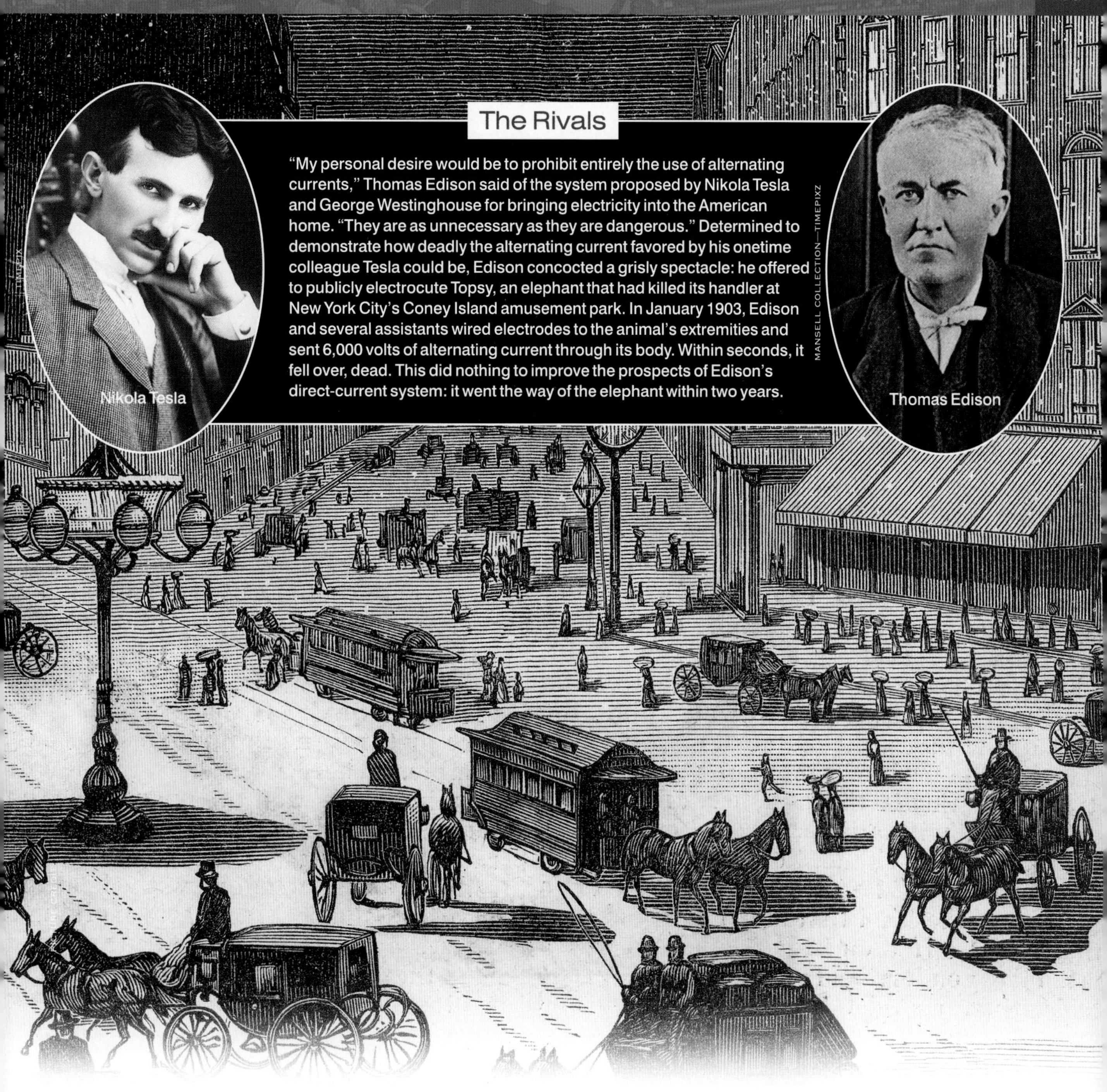

The Rivals

"My personal desire would be to prohibit entirely the use of alternating currents," Thomas Edison said of the system proposed by Nikola Tesla and George Westinghouse for bringing electricity into the American home. "They are as unnecessary as they are dangerous." Determined to demonstrate how deadly the alternating current favored by his onetime colleague Tesla could be, Edison concocted a grisly spectacle: he offered to publicly electrocute Topsy, an elephant that had killed its handler at New York City's Coney Island amusement park. In January 1903, Edison and several assistants wired electrodes to the animal's extremities and sent 6,000 volts of alternating current through its body. Within seconds, it fell over, dead. This did nothing to improve the prospects of Edison's direct-current system: it went the way of the elephant within two years.

house's partner—soon produced "transformers" that could do so. In this way, a regional and even national "grid" of generators and transmission lines could be linked together. It was clearly superior to Edison's localized plan.

The trick was to use alternating current (which flows back and forth over two wires), rather than Edison's direct current (which flowed in a single direction over one line). Tesla also invented the first practical alternating current motor: now electricity could power mechanical devices (like factory machinery or toasters) as well as lights. In 1888, Westinghouse licensed the rights to all Tesla's work with alternating current. The battle was on to power this new Age of (Electric) Enlightenment.

Desperate to showcase the merits of his system, Westinghouse drastically underbid Edison to supply power to the 1893 World's Columbian Exposition in Chicago. In 1896 he set up the world's first hydroelectric generator at Niagara Falls and sent power 25 miles away to Buffalo, N.Y.

By 1901, Buffalo was known as the City of Light and had been chosen to be host of the Pan-American Exposition: the advantages of alternating current were clear. Within four years, Edison surrendered, selling his patents and generating stations to General Electric, a "trust" company formed by financier J.P. Morgan to halt the chaotic competition between differing standards. Soon, the world was wired—in AC. ■

WHEELS WITHIN WHEELS
Tesla poses in front of one his Tesla Coils. He arrived in America with only 4¢ in his pocket, acquired wealth and fame and died with his pockets almost empty again

Profile

Alternating Currents

Everyone knew that Nikola Tesla was charged with visions. But no one knew which Tesla—the genius or the crank—created them

I KNOW TWO GREAT MEN AND YOU ARE ONE OF them," the note read. "The other is this young man." So wrote Charles Batchelor, a director of the Paris-based Continental Edison Co. to the firm's New York City headquarters in 1884, introducing a young Croat, Nikola Tesla to his boss, Thomas Edison. Tesla had studied in Prague and Budapest and was already regarded as a genius at age 28. Edison hired him on the spot. Within a matter of weeks, Tesla assured Edison that he could improve the efficiency of the company's electric generators 25%. A skeptical Edison promised Tesla a bonus of $50,000 if he made good. Tesla quickly delivered on his promise—and Edison quickly reneged on his. "Tesla, you don't understand our American humor," he said, refusing to pay up. Tesla stormed out of Edison's lab, vowing he'd show the world who was the greater genius.

Tesla quickly formed a partnership with railroad innovator George Westinghouse to develop his ideas for alternating current, a vastly more practical way of producing electric power and conveying it over long distances than the direct current Edison favored. Edison marshaled his potent publicity machine in the battle, but the advantages of Tesla's system won out.

Fascinated with radio, Tesla filed patents for a system that, five years ahead of Guglielmo Marconi, would transmit telegraph signals through the air. But he soon lost interest in the project and never got around to building it, a trait that became a maddening pattern. In 1900, he designed (but again, never built) the first workable radar system. These deeds brought him wealth, fame—and a reputation for erratic behavior. A contender for the Nobel Prize in Physics, he took himself out of the running twice, infuriated that rivals Edison and Marconi were considered for the honor.

Tesla also alienated the wealthy backers who were the indispensable patrons of early 20th century inventors. He took $100,000 from John Jacob Astor IV, promising to work on improving fluorescent lighting and vacuum tubes, but quickly abandoned this work for a series of experiments in which he induced violent electrical storms in Colorado and caused massive blackouts. He convinced J.P. Morgan to finance research into a wireless telephone, but spent the money attempting to fire lightning bolts across the Atlantic Ocean and light the 1902 Paris Exposition from New York. Neither plutocrat was amused. When he wasn't baiting and switching millionaires, Tesla was branding Albert Einstein a crackpot: his relativity theory would be shredded—shredded!—when Tesla published his theory of gravitation. He never did.

> "Tesla was a poet," Thomas Edison once declared, calling his rival's ideas "magnificent but utterly impractical"

The line between genius and madness is often blurred; in Tesla's later years, it seems to have disappeared entirely. His obsessions—fear of germs and a fascination with the number three—took over. His funds exhausted, he spent hours each day on the streets of New York City feeding pigeons. He died in 1943, just six months before the U.S. Supreme Court determined that he was the true inventor of radio. His name lives on, if barely: a "tesla" is the basic unit of force in magnetic induction. He is also the hero of a cult that believes he was an alien masquerading as a human being. Shortly before his death, Tesla said of himself and his rivals, "The present is theirs; the future, for which I really worked, is mine." Sadly for Tesla, that future has not yet arrived. ■

Target: Cleaner Power

We live in a digital age—but we're still burning the same fuel as the Model T

We all know where we need to go—it's just taking awhile to get there. To ensure a sustainable future on this planet, we must move from dirty, exhaustible power sources to clean, inexhaustible power sources. Inventor Dean Kamen confides in a profile earlier in this book that he is experimenting with a new source of power: he joins thousands of scientists around the world who are working to bring about an age of clean energy. The search is focused on resources that are constantly renewed by nature: wind power, solar power, hydrogen power, hydroelectric power and geothermal systems that use the earth's heat. A brief review of promising areas:

■ **WIND POWER.** Today, Denmark gets 18% of its energy from the wind, free as a breeze. Windmills are old news, of course, but today's models are more

I'LL FOLLOW THE SUN
In Australia, a battery of shiny dishes made up of photovoltaic cells harnesses the sun's power. Like phototropic plants, the dishes swivel to follow the sun's path across the sky

What About Nuclear?

Soviet scientists built the world's first functioning power plant that harnessed a nuclear reaction to create electricity: the AM-1 reactor at Obninsk, 70 miles southwest of Moscow, went online on June 27, 1954. It generated power for five years, then became a research site. It was shut down for good on May 6, 2002.

Long touted as the fuel of the future, nuclear power suffered two crushing blows in recent decades: the near-meltdown at the Three Mile Island plant in Pennsylvania in 1979, followed by the 1986 breach of the reactor core at the Chernobyl plant in Ukraine. Disposal of radioactive waste remains a huge problem. And don't forget terrorism: a bomb that takes out a wind farm is far less frightening than a bomb that takes out a nuclear plant.

YANN ARTHUS-BERTRAND—CORBIS

ENERGY FROM THE ATOM
Lights illuminate a nuclear reactor core, submerged to control the intense heat it generates

DAVID MCNEW—GETTY IMAGES

THE ANSWER, MY FRIEND ...
The 20 giant fans at the Middelgrunden Wind Farm in the Baltic Sea near Copenhagen, Denmark, provide power for 32,000 homes. The blades are 100 ft. long. Wind power is currently the world's fastest-growing renewable energy source

efficient than ever, thanks to pioneering Danes and Germans who developed better materials, blades and rotors.

One obvious question: What do you do when the wind stops blowing? Scientists hope to store wind power by using it to provide current that extracts hydrogen from water molecules to be used in hydrogen fuel cells, which generate pollution-free electricity.

■ SOLAR POWER. Scientists have been working to draw energy from the sun since French radioactivity pioneer Edmund Becquerel discovered the photovoltaic effect in 1839, proving the sun radiates electric power. In the 1800s, primitive solar collectors were used to power steam engines, but they were very inefficient. As a result, photovoltaic power was put to work only infrequently—in light meters in cameras, for instance.

After U.S. scientist Russell Ohl discovered in 1941 that certain semiconductors are highly reactive to sunlight, a new generation of more efficient collectors emerged. Steadily improving, solar power will most likely find future use in "distributed" form, supplying limited local needs, such as powering and heating homes.

■ HYDROGEN POWER. Energy scientists agree: the goal is to move beyond fossil fuels like oil and coal and begin using hydrogen. Fuel-cells are now working to power automobiles, and are being tested as small generators and for other uses. But before distributed fuel cells can replace today's giant electric plants, an Edison of the hydrogen age must make them more efficient and less costly. The race is on: Gentlemen, start your (hydrogen) engines. ■

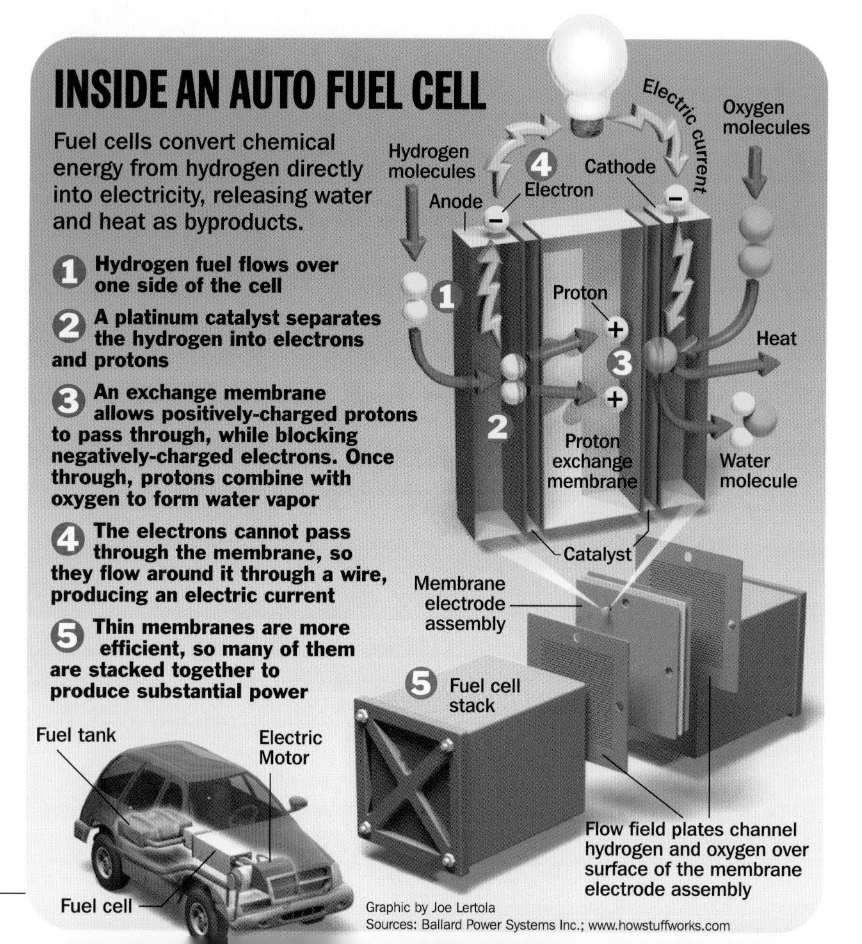

"All Hail, King Steel"

Two innovations made it the indispensable backbone of modern life

Out of the center roared ferocious geysers of saffron and sapphire flame, then a stream of light that flung violet shadows everywhere. A fountain of sparks followed, like 10,000 gorgeous rockets. They fell in a beautiful curve, like the petals of some enormous flower." The writer of this 1893 article in *McClure's* magazine wasn't describing a fireworks display. Rather, he was rhapsodizing about a new process for making steel. If the rhetoric seems overwrought, keep in mind that even if every other ingredient necessary for Stage 2 of the Industrial Revolution—steam engines, long-distance freight transport and electric power—had been in place, it still couldn't have happened without affordable, high-quality steel.

As early as 300 B.C., ancient tribes were mining iron ore from the ground, melting it to remove impurities and then hammering the still-soft metal as it hardened to force out microscopic air bubbles. The result—steel—was far stronger and lighter than iron, but the refining process was slow, expensive and inconsistent.

In 1855, English inventor Henry Bessemer (then known for improving the making of pencils) perfected a new process for blasting air into a furnace where molten iron was being refined into steel. This caused impurities within the metal to bubble to the surface, where they could be washed away. The "blast furnace" became known as the Bessemer Converter; the new process produced large quantities of steel in as little as 20 minutes, at a cost of a few dollars per ton. The well-financed Bessemer adapted a similar process developed a decade earlier by U.S. inventor William Kelly, whose patents he bought for a pittance.

FRANK PEDRICK—THE IMAGE WORKS

The following year, two German-born inventors who had immigrated to Britain patented the first regenerative furnace. William and Friedrich Siemens realized that much of the heat generated in an oven escapes as waste. They developed the first practical way to channel hot gases back into the chamber, where they contributed to superheating the inside of the oven. The Siemens brothers imagined that their invention would be used mainly by glass blowers, but within 10 years it had joined the Bessemer process as an essential tool for steelmakers.

By 1877, Andrew Carnegie had licensed both the Bessemer and Siemens technologies and was on the way to making the first great fortune in American steel. As the era of railroads, bridges and skyscrapers dawned, Carnegie was moved to some flowery rhetoric of his own. On Jan. 1, 1901, when he relinquished control of Carnegie Steel to J.P. Morgan in exchange for $250 million, Carnegie wrote, "Farewell, then, Age of Iron; all hail, King Steel." ■

ON THE GRID
Workers create a wall of rebar as they build a wastewater exchange box

Concrete

For decades after cheap, high-grade steel became ubiquitous, large buildings were still made mostly of stone, restricting their height. Engineers finally adapted a technique first used by Parisian Joseph Monier, who around 1849 began making his garden pots stronger by adding a steel mesh to the clay. Decades later, builders found that concrete reinforced with steel bars (rebar, above) was strong and light enough to support structures of almost unlimited height.

The picture at right shows one of Thomas Edison's periodic flops, an attempt to create cheap housing by casting concrete into prefabricated molds. The "Wizard" even designed concrete beds and concrete pianos for his concrete palaces. But Edison never managed to invent his a way around the chief difficulty that faced his "salvation for the slum dweller": nobody wanted to live in them.

UNDERWOOD & UNDERWOOD—CORBIS

Men of Steel

"I had an immense advantage over many others dealing with the problem," Henry Bessemer later recalled of his invention of the first practical method for mass-producing steel, "inasmuch as I had no fixed ideas derived from long-established practice to control and bias my mind." He wasn't kidding: Bessemer's previous work had been with typography, paint dyes and telescopes. William Siemens was similarly fortunate: his insights into steelmaking came as a result of efforts to improve the efficiency of steam engines by limiting the amount of heat they wasted. Siemens and his brother Werner later went on to patent the first practical dynamo (a crude form of electric generator) and help lay the first transatlantic telegraph cable. Both men were eventually knighted for their contributions to British industry.

MANSELL COLLECTION—TIMEPIX (2)

Henry Bessemer

William Siemens

BROWN BROTHERS

HOT STUFF

A dolomite-lined steel vat pours newly refined molten steel from a gigantic Bessemer Converter into a mold

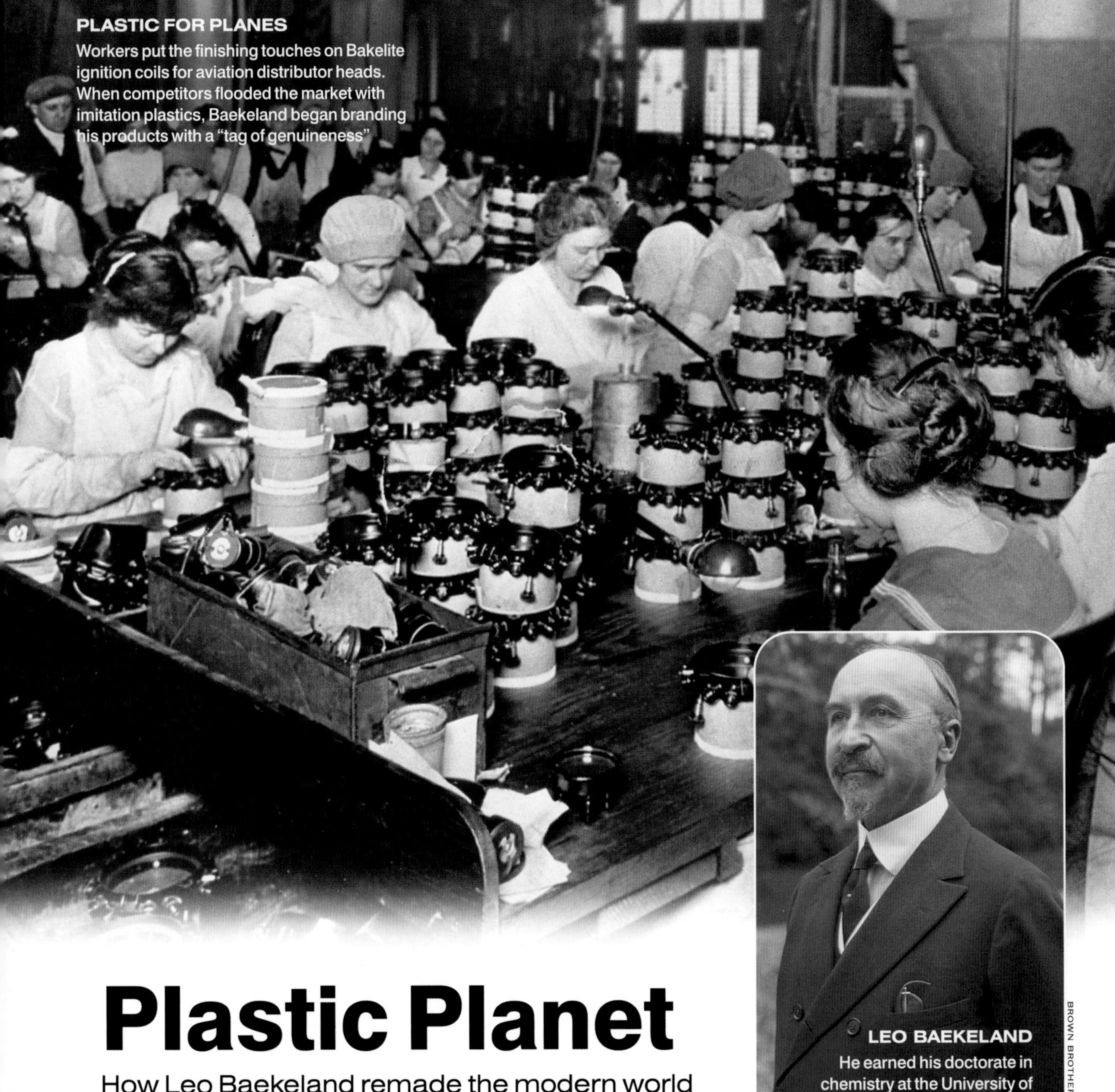

PLASTIC FOR PLANES
Workers put the finishing touches on Bakelite ignition coils for aviation distributor heads. When competitors flooded the market with imitation plastics, Baekeland began branding his products with a "tag of genuineness"

LEO BAEKELAND
He earned his doctorate in chemistry at the University of Ghent in his native Belgium

BROWN BROTHERS

Plastic Planet

How Leo Baekeland remade the modern world

History has its Bronze Age. Andrew Carnegie boasted of founding an Age of Steel. One hundred years later, we think of ourselves as living in the Digital Age or the Age of Information. But down deep, we know the truth: we live in the Plastic Age. And the man who made up the stuff that makes up our world was Leo Baekeland, who sailed to America from Belgium in 1889 harboring every immigrant's dream: to work hard and get rich. And he was well equipped for the task, for he managed to unite within himself the two personas of innovation—the inventor and the entrepreneur—that, when divided between two hard-driving individuals, often spark mutual repulsion and lawsuits. Driven by opportunity as much as curiosity, the chemist smartly staked his claim in emerging technologies and found his first success with Velox, a special paper that developed photographs in artificial light. George Eastman was so impressed he paid Baekeland the then incredible sum of $1 million for it.

Baekeland settled in Yonkers, N.Y., where he built a big barn for research behind his home. Soon he found another market niche in the burgeoning world of electricity, where demand for shellac far exceeded the supply. Shellac is an organic substance, a residue deposited by beetles on trees in southern Asia, then harvested, refined

1950: BAKELITE TV

As cheap imitations flooded the market, plastics were often scorned. But authentic Bakelite products, like this British TV, have a beauty of their own that is increasingly prized by collectors

NMPFT—TOPHAM—THE IMAGEWORKS

and used as a varnish—or, as the world was quickly learning, as a fine insulator of electricity.

So Baekeland set out to create a synthetic shellac. Working with an assistant, he filled three years and scores of lab journals doing the drudge work of chemistry: measuring, mixing, brewing, heating, distilling. He found promise in the goo left behind when formaldehyde (the alcohol-based embalming fluid) reacted with phenol (a turpentine-like solvent distilled from coal tar). When "Doc" Baekeland cooked them up under pressure in his nifty "Bakelizer"—a modern-day alembic that allowed him to control the reaction between the two—he ended up with a transparent substance that was hard yet moldable: plastic.

Enter Baekeland the entrepreneur. He unveiled his miracle matter in 1909 at a meeting of the American Chemical Society. The assembled scientists knew they were peering into the future, at a substance that was sturdy, inexpensive and infinitely malleable. Bakelite could be used to make just about anything, from rosary beads to radios, telephones to billiard balls—and it soon was. It was the world's first completely synthetic plastic (celluloid, which Eastman used earlier to make film, was a cotton-based, organic product). Since it first hit the molds, plastic has molded our world. And Leo Baekeland, even wealthier, spent his later years sailing his beloved yacht, the *Ion.* One word, Doc: plastics. ■

BERNARD ANNEBICQUE—CORBIS SYGMA

Tyvek, Kevlar & Co.

As corporations counted the blessings of Bakelite—inexpensive, malleable, strong, tintable—the race was on to find synthetic replacements for organic materials. Nylon, rayon, vinyl, polyethylene, polyurethane: the polysyllabic parade marches on. Two more recent plastics highlight the versatility of these not-found-in-nature materials.

Tyvek, used extensively as a building insulator, was discovered the old-fashioned way—by pure accident. Du Pont chemist Jim White noticed white polyethylene fluff coming out of a pipe in a lab in 1955. Presto! He had found a tough new plastic that combines the qualities of paper, film and fabric. And, as Du Pont's website boasts, it's vapor-permeable!

Kevlar is also a Du Pont product, discovered by chemists Stephanie Kwolek and Herbert Blades in 1965. A fabric stronger by weight than steel, it is best known as the the tough stuff used in bulletproof vests, and as the heat-resisting protective garb worn by fire fighters, shown in the 1996 picture at top right.

TIM BOYLE—GETTY IMAGES

WORKIN' ON THE LINE
The breakthrough visible in this 1913 picture is the chain that attaches the chassis to the overhead conveyer belt, moving the workload past the worker

Of Men and Machines

Do this. Good job! Now do it again ... and again ... and again ... and again ...

Henry Ford was in a fix. The breakthrough automobile he'd introduced in 1908, the Model T, was a runaway success. Demand for the Tin Lizzies far outpaced his ability to supply them. Although he opened a new plant in Highland Park, outside Detroit, on Jan. 1, 1910, Ford couldn't seem to build the damn things fast enough. Obsessed with reducing the amount of time it took to assemble each car, Ford concentrated on standardizing all his parts, then ordered up new machines from his in-house designers—machines specifically designed to perform only one function with maximum efficiency.

Still, Ford's men took 17 hours to create a Model T, using a three-tiered system. First, bare auto chassis were mounted on sawhorses down the middle of the vast factory. Second, "parts runners" dropped off the next parts needed in the assembly process, just before the third tier arrived—the assemblers, moving down from the last chassis on which they'd performed their specific task, affixing the tires or installing the engine. It was a moving assembly line—but most of the moving was being done by the parts runners and the assemblers.

So Ford worked to eliminate the parts runners, placing specialized machines next to the place on the line where the parts they made were used. Faster? Yes—but still not fast enough to suit Clarence Avery, Ford's production expert. So the runners were replaced by conveyor belts. Faster? Yes—but still not fast enough. Finally, Ford recalled the slaughterhouses he had seen in Chicago, in which animal carcasses were trundled about on overhead tracks. In 1913 a rope-and-winch system was put into the plant, and the flow was reversed: now the chassis, hooked from above, moved past the assemblers. The rope system was quickly replaced by a continuous moving chain. By 1914, Model Ts were rolling off the factory floor in 90 minutes rather than 17 hours.

Ford wasn't the first to discover the value of division of labor, standardized parts and moving assembly lines; rival automaker Ransom Olds used some of these techniques in the early 1900s. But Ford brought the system to perfection, establishing the template for all manufacturing that followed.

Did the automation of labor turn the laborers themselves into automatons? The charge is not a new one: indeed, Czech writer Karel Copek introduced the concept of the robot in his 1921 play *R.U.R.* only eight years after Ford automated his line. *Robota* is the Czech word for "serf" or "drudge worker"—but don't tell R2-D2. ■

Now, Metal Mutts

As automation standardized labor by reducing it to a series of specialized tasks, the notion of the automated man became ever more resonant. Long before scientists and engineers actually began to create robot technology, popular culture was filled with automatons, from Fritz Lang's 1927 dystopian film classic *Metropolis* to countless men of metal—both sidekicks and villains—in comic books and Saturday-afternoon movie serials.

Today robot machines are at work around the world, defusing bombs, vacuuming homes, even exploring Mars in the form of the Sojourner space probe. But as these machines don't resemble humans, they don't fascinate us.

One robot that does ape appearance and behavior is Sony's series of robot dogs, launched in 1999, and increasingly popular as a rich person's toy. "Aibo" looks through a color camera; feels with heat sensors, an infrared range finder and touch sensors; moves via acceleration and speed sensors; and speaks through a stereo microphone. A 2003 model fetches $1,300.

KOICHI KAMOSHIDA—NEWSMAKERS—GETTY

TURNING RATS INTO ROBOTS

Scientists in Brooklyn, N.Y., are wiring rats to perform dangerous tasks, like searching for human bodies after fires. Using classic behavioral conditioning, researchers implant electrodes in clumps of rat-brain cells that govern whisker sensation and pleasure. When human controllers want the rat to turn left, they beam a signal that triggers a mild electric shock to the left-whisker cells. If the rat responds by turning left, it's rewarded with stimulation of the pleasure cells. A right turn is signaled by a pulse to the right-whisker cells.

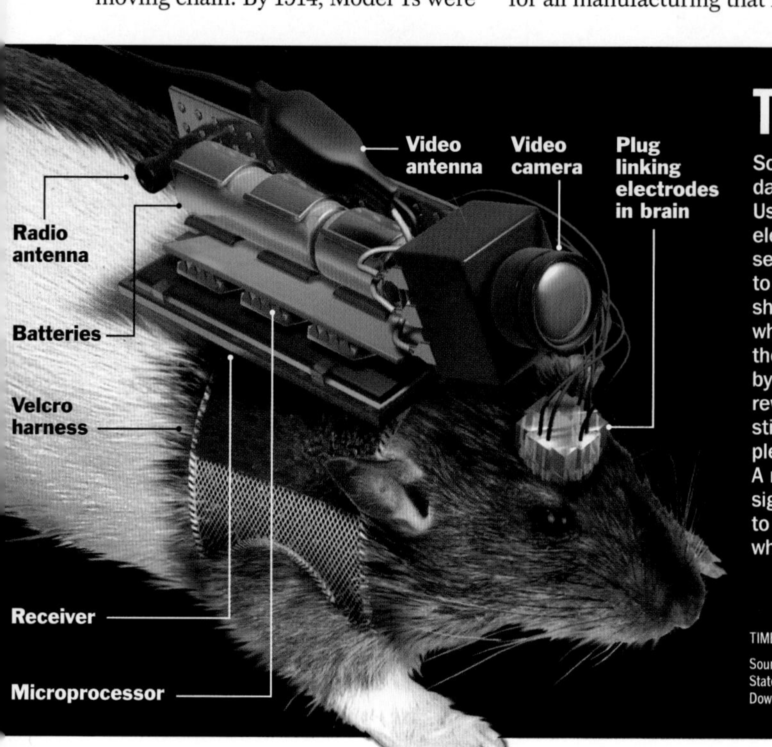

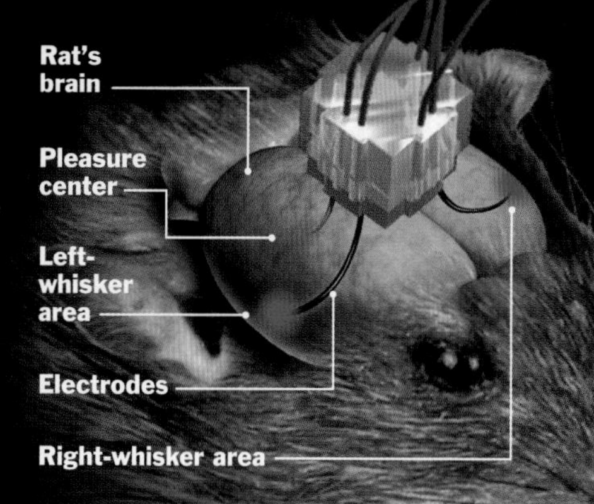

TIME Diagram by Ed Gabel

Source: Sanjiv Talwar, State University of New York Downstate Medical Center

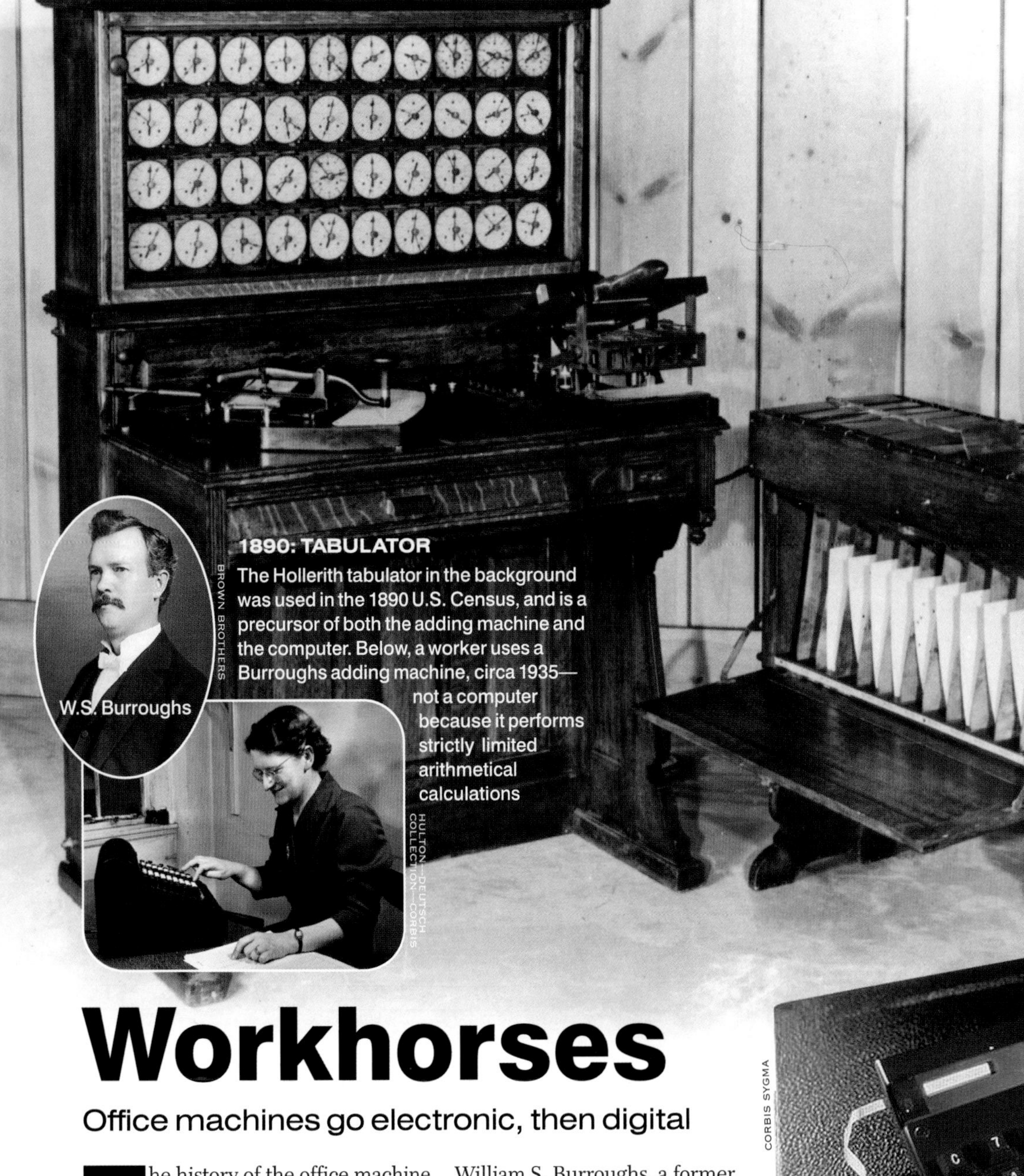

FRANK BAUER—TIMEPIX

HULTON ARCHIVE—GETTY

BROWN BROTHERS

1890: TABULATOR
The Hollerith tabulator in the background was used in the 1890 U.S. Census, and is a precursor of both the adding machine and the computer. Below, a worker uses a Burroughs adding machine, circa 1935—not a computer because it performs strictly limited arithmetical calculations

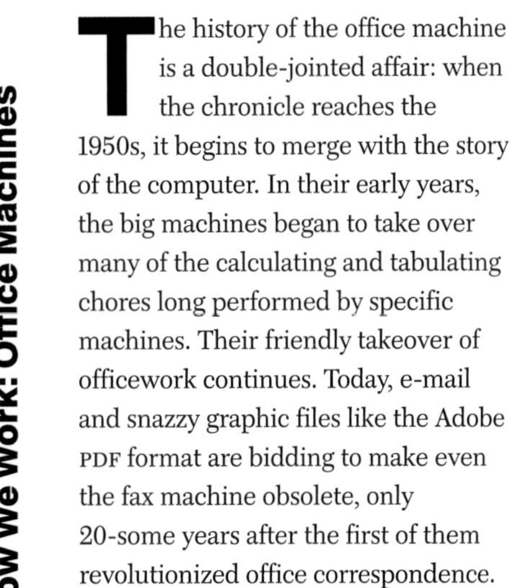

HULTON-DEUTSCH COLLECTION—CORBIS

Workhorses

Office machines go electronic, then digital

The history of the office machine is a double-jointed affair: when the chronicle reaches the 1950s, it begins to merge with the story of the computer. In their early years, the big machines began to take over many of the calculating and tabulating chores long performed by specific machines. Their friendly takeover of officework continues. Today, e-mail and snazzy graphic files like the Adobe PDF format are bidding to make even the fax machine obsolete, only 20-some years after the first of them revolutionized office correspondence.

The adding machine, born in 1885, had a good long run. Its inventor was William S. Burroughs, a former bank clerk from New York State who was working for the Boyer Machine Co. in St. Louis. He was intrigued by the challenge of building a better calculator: early models were hand-cranked, and each finicky unit worked only if yanked just the right way. Burroughs invented a "dash pot," an oil-filled hydraulic governor that regulated the crank, producing a correct sum every time. Although Burroughs died young, his company came to dominate its field—and his name entered American literature through the avant-garde novels of his eponymous beatnik grandson. ■

CORBIS SYGMA

1967: PORTABLE CALCULATOR
Move over, Mr. Burroughs. As part of general downsizing sparked by the transistor revolution, Texas Instruments inventors Jack St. Clair Kilby, Jerry Merryman and James Van Tassel cooked up the first electronic handheld calculator in 1967. Kilby is a pivotal figure in the transistor revolution; he shares the credit with Robert Noyce and Gordon Moore for inventing the integrated circuit

1949: PHOTOCOPIER

Older readers can recall the days when making copies involved mimeograph machines or carbon paper: either way, it was a smudgy, foul experience. The man who made copying painless was U.S. inventor Chester Carlson, above, who invented the first photocopy machine in 1939, with an assist from physicist Otto Kornei. It took Carlson five years to find a manufacturer to market his device; which he called xerography, from the Greek *xeros* (dry) and *graphein* (to write). As the Haloid Co. grew, it first became the Haloid Xerox Corp. and eventually Xerox

DAVID E. SCHERMAN—TIMEPIX

1873: STOCK TICKER

Technology has not always embraced streamlined design. This beautiful stock-ticker transmitter was invented by Thomas Edison—as its label boasts. Edison was a former telegraph operator; this machine incorporates innovations made by others over two decades

Binders & Reminders

The humble paper clip is emblematic of the spirit of invention. Yes, it appears to be simple at first glance—just a straight length of wire bent into a very specific shape. But the classic at right, the Gem #1, capped a period of more than 30 years during which various inventors tried to create clasps to replace the ribbons and straight pins that bound together 19th century papers. Yet England's Gem Manufacturing Ltd. never patented its double-oval design. Intrigued? Henry Petroski, the gifted historian of technology, brings the fascinating history of the paper clip alive in his book *The Evolution of Useful Things.*

The Post-It note was invented by a 3M engineer, Art Fry, who based it on a gentle adhesive discovered by a colleague, Sheldon Silver. Fry was divinely inspired: he was looking for reusable bookmarks to replace the slips of paper that kept falling out of his church choir books.

1977: POST-IT NOTES

The product didn't catch on right away. Manufacturer 3M stuck to it, and now the yeller fellers are ubiquitous

PHILLIP JAMES CORWIN—CORBIS

Attention, Shoppers!

One man reinvented the grocery store by setting the customers free

1881: CASH REGISTER

Retail's workhorse was invented by James Ritty and John Birch in 1881. Their patent was bought by John C. Patterson, who improved the machine by adding the paper roll, founded the National Cash Register Co. and offered the model at left for sale in 1885

BETTMANN CORBIS

John C. Patterson

CULVER PICTURES

America's calendar of holidays includes Father's Day, Mother's Day, Presidents' Day and other annual celebrations whose main function seems to be to promote retail sales. Fair enough. But should we ever level with ourselves and establish a holiday to celebrate what we really love—shopping—there would be no question as to when to observe it: Sept. 16.

On this date, in 1916, an inspired entrepreneur, Clarence Saunders, threw open the doors to a new grocery store on Jefferson Street in Memphis, Tenn. He had given it an unlikely moniker: Piggly Wiggly. But the real news was inside, where Saunders had rethought the relationship between grocer and customer. Previously, shoppers had arrived at a store and presented a list of needs to a clerk standing behind a large counter, who then retrieved the items. There was no need to discuss the customer's preference; after all, flour was flour, a commodity bought in bulk by the grocer and measured out to the consumer.

Not at Piggly Wiggly. Here customers were handed a shopping basket and invited to serve themselves. They wandered down narrow aisles whose shelves were packed tight with goods, including a selection of branded products, that they placed in their basket. Their only contact with store personnel occurred at the end of their journey, when they handed their purchases to a clerk and were billed.

The process was revolutionary. Saunders' hunch was correct: when he allowed the customers to shop for themselves, his sales skyrocketed, for the newly footloose patrons were now free to buy on impulse.

But the implications of self-service shopping were even more wide-ranging. The practice, which swept the nation after its introduction in Memphis, encouraged national branding and advertising. It focused attention on a product's packaging. And it required each item to be marked for price, sparking price battles among suppliers. And as self-service shopping caught on, markets began stocking more and more goods. These "supermarkets" began to compete directly with the one-man, one-product purveyors that used to make up Main Street, butchers and bakers and fishmongers.

Saunders became a millionaire, but he lost his empire in a series of Wall Street stock shuffles. And why did he call his store Piggly Wiggly? "Just so you'd ask me that question," the Barnum of Bread used to say. ■

CORBIS

1879: IVORY SOAP

Candlemaker William Procter and soapmaker James Gamble joined forces in 1837. Their company helped establish the concept of the "brand," a maker's mark that promised uniform quality. One early national brand: a floating soap, Ivory

1916: SELF-SERVICE GROCERY
Lunatics in charge of the asylum, newly liberated shoppers pack Saunders' first Piggly Wiggly store in 1918. After he lost control of the chain, Saunders worked on a prototype of a fully automated grocery store, the Keedoozle ("Key Does All"), but the project failed

CORBIS

The Wonderful Wishing Machine

In the late 19th century, for the some 65% of Americans who lived in rural towns, the Sears-Roebuck catalog *was* America. Here, beneath a beautifully illustrated cover, was the nation's promise of abundance, a carefully classified cornucopia of goods, all available via mail. The famed catalog invented a new paradigm in buying and selling, direct-mail marketing. Its creator was Richard Sears, a railroad agent in Minnesota. In 1886, when a shipment of watches he received was not wanted by the local jeweler, Sears bought them and sold them to train riders. The process sparked an idea: Why not sell goods directly to consumers, delivering them via mail, and thus avoid the shopkeeper middleman, with his notoriously high markup of prices? Sears soon moved his business to Chicago; Alvah Roebuck answered his ad for a watchmaker and became his partner. By 1895, only a few years after its launch, their catalog had 532 pages.

THE GRANGER COLLECTION

BETTMANN/CORBIS

How We Play

Shaler's Flexible Roller Skate, patent drawing, 1861

9

1927: CYCLONE
Coney Island's famed Cyclone is not the first roller coaster. At 85 ft., it's not the tallest either. But its slant (60 degrees), speed (more than 60 m.p.h.) and sound (from creaking wooden timbers) rank it among the scariest. Now listed on the National Register of Historic Places, the Cyclone was described by Charles Lindbergh in 1927 as "more frightening than flying an airplane at top speed"

Merry, Scary Machines

It's a tough job, but somebody had to sit down and dream up the bumper car

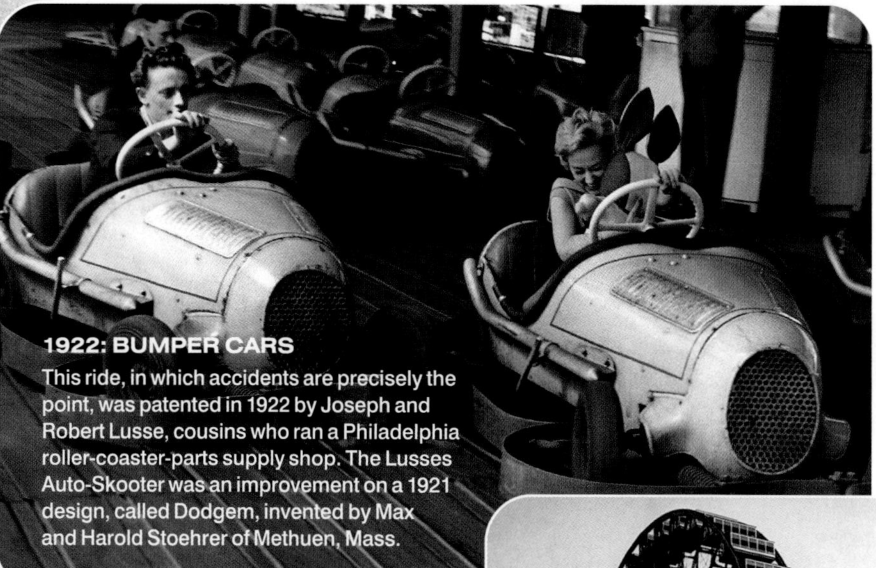

HULTON DEUTSCH COLLECTION-CORBIS

1922: BUMPER CARS
This ride, in which accidents are precisely the point, was patented in 1922 by Joseph and Robert Lusse, cousins who ran a Philadelphia roller-coaster-parts supply shop. The Lusses Auto-Skooter was an improvement on a 1921 design, called Dodgem, invented by Max and Harold Stoehrer of Methuen, Mass.

In the months leading up to the Chicago Exposition of 1893, its promoters were desperate to find an "original, daring and unique" attraction for their fair. Specifically, they wanted to trump the the Eiffel Tower, the star of the Paris Exhibition four years earlier. But the proposal of a 33-year-old civil engineer, George Ferris, for a 264-ft.-high "observation wheel" was a bit too original and daring. They relented only after other engineers approved the design. The amusement wheel was not entirely new (earlier models were made of wood and didn't top 50 ft.), but Ferris' steel wheel was a revelation, by several orders of magnitude. Each of his wheel's 36 cars held 60 people. While two 1,000-h.p. steam engines gave the wheel its spin, a gigantic airbrake added stopping power.

Ferris' wheel was a smash: 1,453,611 customers ponied up 50¢ each to view bustling Chicago from on high. But Ferris didn't profit: after having to litigate with the fair's management over his fee, he died broke in 1896.

While Ferris was fighting with his partners, John A. Miller ("the Thomas Edison of the Roller Coaster") was fighting gravity. Among Miller's 100-plus roller-coaster patents was a device that locked the cars to the tracks, while his "safety ratchet," which prevents cars from sliding back down an incline by snapping onto latches every few feet, gave the coaster its distinctive clanking sound. In the process, Miller turned what the Patent Office once called "pleasure railways" into modern scream machines.

A key invention remained. Walt Disney's 1955 masterpiece, Disneyland, transformed the sleazy old midway of freak shows and arcades into a squeaky-clean family "theme park" that has now entertained 400 million visitors. Thanks to Walt, the midway is living happily ever after. ■

BROWN BROTHERS

1893: FERRIS WHEEL
The public wouldn't embrace Ferris' contraption until he proved it could withstand the high winds from Lake Michigan, so he ascended to the top with his wife in a 90-m.p.h. gale. Modeled on a bicycle tire, the big wheel was built in four months

ALLAN GRANT-TIMEPIX

1955: DISNEYLAND
Ronald Reagan was an honorary master of ceremonies as more than 30,000 guests crowded into Disney's Anaheim, Cal. park on opening day, right. The larger Walt Disney World opened 16 years later in Florida

Fun and Games

Once upon a time, kids had no Teddy Bears to hold

Toys are inventions too, even if we don't usually think of them that way. And from the yo-yo (once used as a lethal weapon in the Philippines) to the Slinky (conceived by a naval engineer who dropped a tension coil and watched it shake like a bowlful of jelly), most of our beloved childhood playthings have a fascinating story to tell. We're only sorry we couldn't tell more: missing here are Erector sets and Tinker Toys, Lincoln Logs and Super-Soakers, Legos and Rubik's Cube—and crayons! ■

1958: HULA HOOPS

Undeniably fun—and undeniably given a perfect name, the hula hoop turned somnolent Americans of the late Eisenhower years into instant Elvises. The plastic fantastic bands were invented by Richard Knerr and Arthur (Spud) Melin, the Wham-o toy company geniuses who also gave us the Frisbee

BETTMANN CORBIS

1929: YO-YO

This ancient toy was a weapon in the Philippines; its name means "come back" in Tagalog. It was brought to the U.S. by a Filipino immigrant in the 1920s; the rights were bought by master impresario D.F. Duncan Sr., who added the slip-string feature and made it a fad

1945: SLINKY

This shimmering whimsy—80 ft. of metal—was invented by Richard James, a naval engineer who was working with tension springs while trying to design a meter to monitor nautical horsepower. The name is from a Swedish word meaning sinuous

BETTMANN CORBIS

1949: SILLY PUTTY

First called "nutty putty," the weird stuff was discovered by GE scientist James Wright. The recipe: silicone oil and boric acid. Sadly, today's less smudgy printing inks stop it from copying the Sunday funny pages as well as it used to

1958: FRISBEE

Yale grads say the flying disc was invented by 19th century Elis who hurled pie tins made by the nearby Frisbie Baking Co. Maybe. The modern version derives from Walter Morrison's 1955 Pluto Platter, bought and adapted by Wham-o

KURT WITTMAN-CORBIS

1959: BARBIE DOLL

Barbie was invented by Ruth Handler, co-founder of Mattel Toys, and named for her daughter Barbara. The first doll to show real—albeit highly exaggerated—adult female attributes, Barbie is often labeled a bad role model, yet she's undeniably beloved by girls everywhere

ALLAN GRANT-TIMEPIX

1960: ETCH-A-SKETCH

Developed by France's Arthur Granjean, this dual-knobbed beauty was first known as L'Ecran Magique. British kids call it the DoodleMaster Magic Screen. What's in the secret sand inside? Aluminum dust and teeny plastic beads

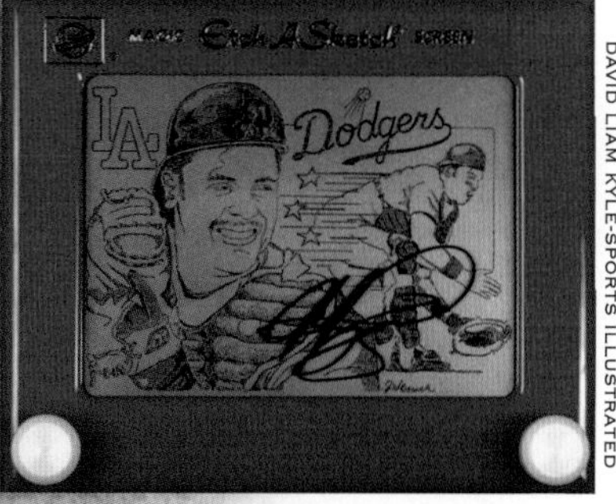

DAVID LIAM KYLE-SPORTS ILLUSTRATED

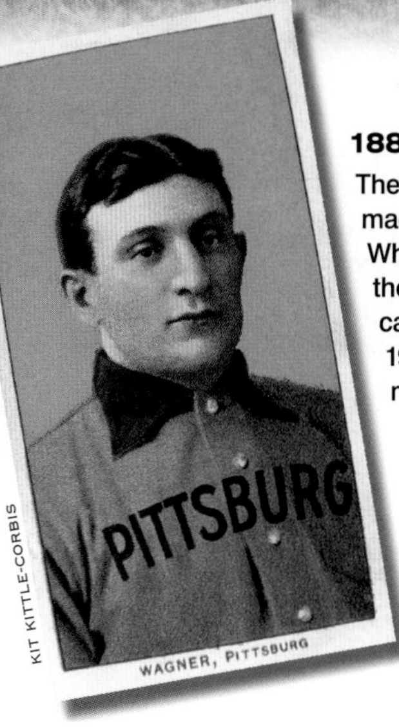

KIT KITTLE-CORBIS

1887: BASEBALL CARD

The first nationally distributed cards were made by Goodwin & Co. of New York City. When cigarette makers began inserting them free in packs, they took off. Later, cards tagged along with bubble gum. This 1909 Honus Wagner classic sold for $1.2 million in 2000

THE GRANGER COLLECTION

1903: TEDDY BEAR

You know this story: after avid hunter Teddy Roosevelt refused to shoot an exhausted, lassoed brown bear, Brooklyn novelty-store owner Morris Michtom put up for sale two bears handmade by his wife Rose, and a fad was born. Germany's cuddly Steiff bear appeared at roughly the same time: a coincidence

Gravity Grooves

Flipping and flying, slipping and sliding—we just can't help inventing ways to mess with momentum

March 1779: three months after Captain James Cook was murdered by Hawaiian islanders, his successor, Captain James King, recorded an observation in the ship's log of H.M.S. *Resolution*, now under his command: "Whenever … the impetuosity of the surf is increased to its utmost heights … twenty or thirty of the natives, taking each a long narrow board, rounded at the ends, set out together from the shore … As soon as they have gained the smooth water beyond the surf, they lay themselves at length on their board, and prepare for their return … Their first object is to place themselves on the summit of the largest surge, by which they are driven along with amazing rapidity toward the shore."

Surfing's roots may be ancient, but the sport was way too much fun for the Yankee missionaries who settled in Hawaii in the 19th century: they banned it. Kept alive by a few traditionalists, surfing was reintroduced to the public in 1905 by teenager Duke Kahanamoku, who rode the waves on 16-ft. "longboards" at Waikiki Beach. Two years later, islander George Freeth gave a surfing demonstration to an enthralled crowd at Redondo Beach outside Los Angeles, using a shorter, 8 ft. board, which was easier to maneuver. Surfing "caught a wave," and it's still … tubular. ■

1935: TRAMPOLINE The modern tramp was developed in the 1930s by two University of Iowa gymnasts, George Nissen and Larry Griswold

GEORGE SKADDING-TIMEPIX

1919: POGO STICK A monster fad in the 1920s, the pogo stick was patented by Illinois furniture maker George Hansburg, who claimed it was based on a hand-crafted model he'd encountered in Burma. Whatever, George!

ROBERT Y. ONO-CORBIS

1965: WINDSURFER The perfect union of surfing and sailing, the Windsurfer was invented in 1965 by two Californians, aeronautical engineer Jim Drake and entrepreneur Hoyle Schweitzer. Sadly, the idea had also been developed independently—and earlier—by Pennsylvanian S. Newman Darby. Years of patent trials ensued. No matter: it's still a great invention

1959: SKATEBOARD

Derived from handmade scooters, the clay-wheeled Roller Derby Skateboard went on sale in 1959. A huge fad in the early 1960s, skateboarding faded as poor design caused too many injuries. Today's better-designed boards (and better protective gear) have made the sport a hit again

BETTMANN CORBIS

CULVER PICTURES

1863: ROLLER SKATES

The wheeled shoes go way back: Briton Joseph Merlin wore a pair to a party in 1760. But in 1863, Massachusetts businessman James Plimpton patented a "rocking skate" that let people make turns and skate backward, igniting the sport's first heyday. The development of advanced in-line skates in the 1980s brought the old sport up to date

1951: FENDER ELECTRIC GUITAR

The electric guitar had several fathers. Jazzman Charlie Christian played an early amplified model; Les Paul hand-built his own pioneering electric, "the Log." California inventor Leo Fender's 1951 solid-body Broadcaster guitar was the template for most future models, like the classic Fender Telecaster above

TED THAI-TIMEPIX

"Music from the Ether"

Roll over, Beethoven: instruments and electronics wed in secret ceremony!

1964: MOOG SYNTHESIZER

Robert Moog at the keyboard in 1986

BETTMANN CORBIS

Musical instruments haven't changed much in centuries: this is part of their glory. But the advent of the electronic age gave musicians and inventors a chance to create new palettes of sound in new ways. The amplification of volume opened fresh avenues all by itself: Bing Crosby's crooning style wouldn't have been possible without the microphone, and it's hard to imagine Chuck Berry duckwalking across a stage playing an unamplified acoustic guitar.

Yet most musical innovators, like Leo Fender and Laurens Hammond, focused on adapting classic instruments to an electronic age, rather than on creating new ones.

The one entirely new instrument that stands out is the theremin, named for the visionary Russian physicist Leon Theremin, who hoped his invention would usher in a new age of music. Although the theremin's unearthly sound created a splash in the 1920s, it survives only as a novelty.

The device is entirely electronic, and here's how it works: magnetic coils use alternating current to generate very weak electromagnetic waves around a small vertical antenna. The player moves his hand within this field, changing its capacity. This alters the frequency of the current and thus changes the pitch of the tones it emits. A separate, ring-shaped antenna regulates volume.

Well, that's the science—but what the description can't convey is the weirdness of the theremin's electronic sounds: notes slide and wail up and down the scale like a banshee in full cry.

This entirely artificial sound found no real counterpart until the 1960s, when Robert Moog, a onetime theremin builder, created the first electronic synthesizer, which was played on a keyboard. A host of copycats followed. In the 1980s the MIDI format (Musical Instrument Digital Interface) ushered musicmaking into the computer age. ■

1934: HAMMOND ORGAN

Laurens Hammond, left, a watchmaker and engineer, created a "tonewheel generator" that altered the frequency of an electromagnet to imitate the sound of an organ

BETTMANN CORBIS

BETTMANN CORBIS

HULTON DEUTSCH COLLECTION-CORBIS

1919: THEREMIN

Leon Theremin plays his instrument, right; he described its sound as "music from the ether." Above, Lucie Bigelow Rosen was a prominent theremin performer in the 1920s, when the device attracted real interest. Seldom heard today, it's most familiar as the eerie woo-eee-ooo-eee sound in the Beach Boys' classic *Good Vibrations*. Led Zeppelin guitarist Jimmy Page also played a theremin, and he still uses one in concert appearances with singer Robert Plant

Pastimes of the Parlor

Do not pass GO if you only claimed to invent Monopoly. Go directly to JAIL!

With the nation in the grip of the Great Depression in the 1930s, Americans had too little of one thing, money, and too much of another, free time. Not surprisingly, it was the great age of board games, and the greatest game was Monopoly.

If you go to the official website, *monopoly.com*, you'll read a charming story about the origin of the famous real-estate game. Hasbro Inc., which acquired the rights to Monopoly in 1991, when it bought original maker Parker Brothers, offers this official history: Monopoly was invented by Charles Darrow, an unemployed gent from Germantown, Penn., who recalled boyhood summers in Atlantic City, N.J., in crafting the board. Darrow brought the game to Parker Brothers in 1934, but their executives weren't impressed with it, noting 52 design errors. But Darrow persevered and paid to have 500 sets made up, which he sold to toy stores and handed out to friends. People loved the game; Parker Brothers decided to buy the rights—and it took America by storm.

AL FRENI-TIMEPIX

1935: MONOPOLY
Charles Darrow, below, was the first game inventor to make more than $1 million from his creation. Parker Brothers went to great lengths to conceal the fact that Darrow's game had several predecessors

BETTMANN CORBIS

Well, the last phrase is correct: Monopoly was a huge fad in the 1930s and remains beloved. But the official history ignores some uncomfortable facts. In 1904, Lizzie J. Magie, a Virginia Quaker, received a patent for a board game called the Landlord's Game. Progressive-era propaganda, it was intended to demonstrate the flaws of capitalism and rental property. Some folks called it Monopoly.

A Williams College student, Dan Layman, played Magie's game in the 1920s and released a version he called Finance. Ruth Hoskins of Indianapolis, Ind., taught the game to friends when she moved to Atlantic City. They devised a version with local street names in 1930 and showed it to a friend from Pennsylvania, who showed it to his friend—one Charles Darrow. After the game became a hit, Parker Brothers quickly, quietly bought the rights to all previous versions, thus acquiring ... a monopoly on Monopoly. ■

TOBY TALBOT-AP/WIDE WORLD

From Pinball Parlor to Video Arcade

ATARI

©LOUIS K. MEISEL GALLERY-CORBIS

1931: PINBALL GAME
Like Monopoly, they are Depression-era kids. The first such games, Bingo, Baffle Ball and Ballyhoo, were introduced in the 1930s; Raymond Moloney invented the first coin-operated game, Ballyhoo, and later named his company, Bally, after it.

The "tilt" mechanism was introduced in 1934, and the bumper three years later. But these early versions lacked the feature that puts the player in charge: the modern-day flipper came along in 1947

1948: SCRABBLE

"The emergence of Scrabble has been volcanic," TIME noted in its July 20, 1953, issue. The game was invented by New York City architect Alfred M. Butts, right, who made 500 sets for friends; he first called his creation Lexico, then Criss-Cross Words. In 1948, social worker James Brunot joined Butts, christened the game Scrabble and began selling it in earnest. By 1953 the game had swept the nation

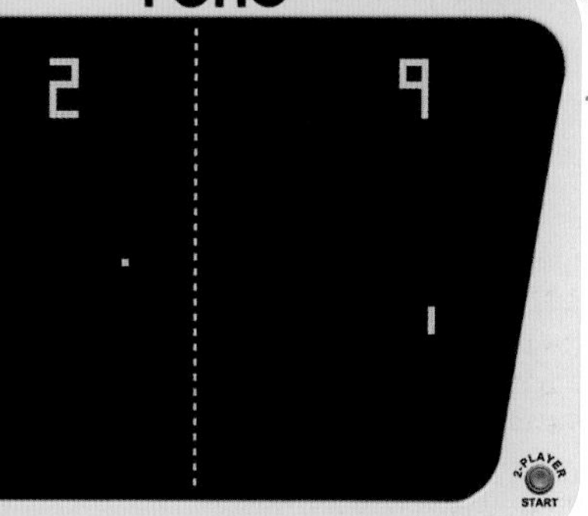

1972: PONG

Steve Russell, a Massachusetts Institute of Technology student, created a game called *SpaceWar* in 1962; it only played on giant mainframes. Utah student Nolan Bushnell played it and went on to create *Pong,* the first coin-operated video game. The first model debuted in a California bar

1983: SUPER MARIO BROTHERS

When Japan's Nintendo game company released *Donkey Kong,* designed by Shigeru Miyamoto in 1981, it was the first arcade video game to feature human characters. Mario the plumber has remained the hero of successive generations of Nintendo games, including the *Super Mario* game, below—which looks dated by 2003 standards

NINTENDO

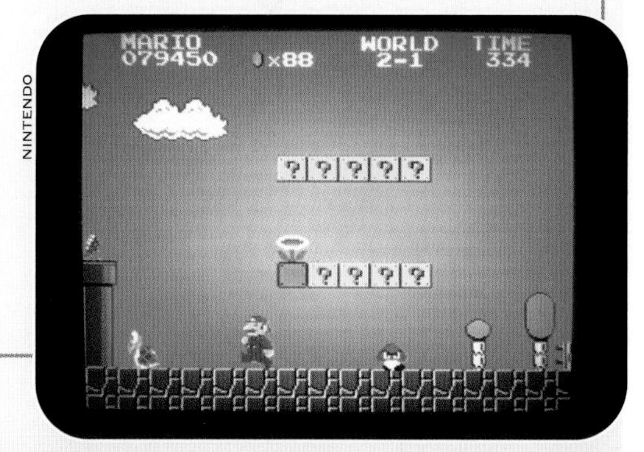

ROYALTY FREE-CORBIS

1913: CROSSWORD PUZZLE

Word grids were known in ancient Rome; in 1913, Liverpudlian Arthur Wynne created a "word-cross" for the Sunday New York *World* newspaper. When a collection of them was published in 1924, the puzzles, with name reversed, became a fad

Profile

Picasso of the Pixel

An artist and an inventor, Shigeru Miyamoto has changed the way the world plays, by giving us fascinating new worlds to play with

SOMETIMES THEY COME TO HIM IN THE BATHtub, where Shigeru Miyamoto works himself into a state of "spiritual tension," the better to open his mind to new ideas. Or the flashes of inspiration may strike while he's working in his garden in Kyoto, Japan. Wherever they begin, Miyamoto's visions end up as pixelated characters—Donkey Kong, the Mario Brothers—that flicker into life on millions of video screens around the world. His fans debate, Is he a modern-day Michelangelo or an updated Thomas Edison? Pick your analogy: either way, the inventor of the modern video game has garnered both fame (pilgrimages from Western pop culture demigods like Paul McCartney, Steven Spielberg and George Lucas) and fortune (his games have earned as much as their music and films, some $6 billion since 1981).

His work has earned more than $6 billion for Nintendo, but he still takes home "a middle manager's salary"

It all began with *Donkey Kong,* which Miyamoto believed (thanks to a creative translation) meant "stubborn gorilla" in English. In 1980, Nintendo, a Japanese playing-card company, asked Miyamoto, an industrial designer who had never got over a childhood obsession with comic books and puppets, to create a new arcade game. Within a year, he came up with a witty, engaging adventure featuring an Italian plumber named Mario whose pet gorilla falls in love with and kidnaps his girlfriend. The first video game to feature recognizable characters, *Donkey Kong* is the daddy of every game face that followed, from *Sonic the Hedgehog* to *Lara Croft.* Suddenly, plain old *Pong* had personality.

Donkey Kong was a huge hit, the biggest-selling video game ever. So Nintendo asked Miyamoto to follow up with a new effort for its new home gaming console. In 1983, Miyamoto designed *Super Mario Brothers,* inventing the side-scrolling format that allowed the game's characters to inhabit a seemingly infinite universe, rather than the static, top-down world of all previous games. *Super Mario's* endless unfolding passageways, he says, were inspired by memories of the sliding doors in his parents' home; the inky blackness of its subterranean world recalls the cool air of the Kyoto caves he explored as a boy.

Super Mario also marked the invention of the "platform" game, in which players progress to new levels that contain entirely fresh backgrounds, experiences and challenges. Through the late 1980s and early '90s, as Nintendo's game systems amped up their computing power, the graphics became ever more sophisticated and realistic, veering away from comic-book images and approaching the quality of movies. In 1997, Miyamoto's *The Legend of Zelda: Ocarina of Time* introduced both space and time to video games: its 3-D format offered a constantly shifting point of view, while the sky shaded from rosy dawn to subtle twilight and then into a star-spangled nightscape.

For all the success he has created, Miyamoto's life remains largely unchanged. He still rides a mountain bike to work each day, still takes home what he calls "a middle manager's salary"—he gets not a penny of royalties from the Mount Fuji of yen his work has earned for Nintendo. He still works in a small office, and he still finds inspiration in unusual places: 2001's *Pikman* was inspired by the communication and cooperation that Miyamoto observed among ants in his garden. His goal, he says, is "just to surprise the world once every few years ... Nintendo allows me to create. I do not need anything other than that." ■

JOHN T. BARR—NINTENDO—GETTY

MASTER CLASS
Miyamoto joins young fans at an arcade. Like many parents, he strictly limits the amount of time his two children can spend playing video games: "Two hours a day—unless it's raining outside, and then I let them play longer"

Index

Index